Nanocomposites in Electrochemical Sensors

Nanocomposites in Electrochemical Sensors

Samira Bagheri

Nanotechnology & Catalysis Research Centre (NANOCAT),
University Malaya, Kuala Lumpur, Malaysia

Iraj Sadegh Amiri

Photonics Research Centre, University of Malaya, Kuala Lumpur,
Malaysia

Amin Termeh Yousefi

ChECA IKohza, Department of Environmental & Green Technology
(EGT), Malaysia Japan International Institute of Technology (MJIIT),
University Technology Malaysia (UTM), Kuala Lumpur, Malaysia

Sharifah Bee Abd Hamid

Nanotechnology & Catalysis Research Centre (NANOCAT),
University Malaya, Kuala Lumpur, Malaysia

CRC Press
Taylor & Francis Group
Boca Raton London New York

CRC Press is an imprint of the
Taylor & Francis Group, an **informa** business

A BALKEMA BOOK

CRC Press
Taylor & Francis Group
6000 Broken Sound Parkway NW, Suite 300
Boca Raton, FL 33487-2742

First issued in paperback 2019

Typeset by MPS Limited, Chennai, India

No claim to original U.S. Government works

ISBN-13: 978-1-138-62677-5 (hbk)
ISBN-13: 978-0-367-88744-5 (pbk)

Library of Congress Cataloging-in-Publication Data

**Visit the Taylor & Francis Web site at
http://www.taylorandfrancis.com**

**and the CRC Press Web site at
http://www.crcpress.com**

Table of contents

Preface   ix
List of figures   xi
List of tables   xv
List of equations   xvii
List of symbols and abbreviations   xix

1 Introduction of nanocomposites in electrochemical sensors   **1**
  1.1   Introduction to nanotechnology   1
  1.2   Application of nanotechnology in catalysis   2
  1.3   Nanomaterials   3
  1.4   Synthesis of nanomaterials   3
  1.5   Nanoparticles   5
  1.6   Synthesis of nanoparticles   5
      1.6.1   Sol-gel method   6
  1.7   Application of nanoparticles in electrochemical sensors 8
  1.8   Chemically Modified Electrodes (CMEs)   9
  1.9   Carbon nanotubes   9
      1.9.1   Electrochemical properties of carbon nanotubes   11
      1.9.2   Purification of carbon nanotubes   11
  1.10 Ferri/ferrocyanide   12
  1.11 Polymers   12
      1.11.1 Polycaprolactone (PCL)   12
      1.11.2 Polyacrylonitrile (PAN)   13
      1.11.3 Chitosan   13
      1.11.4 Gelatin   14
  1.12 Analytes   15
      1.12.1 Carbohydrates   15
          1.12.1.1 Glucose   16
          1.12.1.2 Sucrose   16
          1.12.1.3 Fructose 16
          1.12.1.4 Sorbitol 16
      1.12.2 *D*-Penicillamine   17
      1.12.3 *L*-Tryptophan   17
  1.13 Immobilization techniques   17

1.14	Electrochemical methods	18
	1.14.1 Cyclic Voltammetry (CV)	18
	1.14.2 Electrochemical Impedance Spectroscopy (EIS)	20
	1.14.3 Tafel equation and electron transfer coefficient	21
	1.14.4 Butler–Volmer equation	22
	1.14.5 Mass transfer control and apparent diffusion coefficient	22
1.15	Aim and objectives	23
1.16	Book structure	23
2	**Literature review of nanocomposites in electrochemical sensors**	**25**
2.1	Sensors	25
2.2	Chemical sensors	26
	2.2.1 Potentiometric sensors	26
	2.2.2 Amperometric sensors	27
	2.2.3 Conductometric sensors	28
2.3	Chemically Modified Electrodes (CMEs)	28
2.4	Carbon nanotubes	29
2.5	Electrodes	30
2.6	Electrochemical methods	30
	2.6.1 Cyclic Voltammetry (CV)	30
	2.6.2 Chronoamperometry	31
	2.6.3 Electrochemical Impedance Spectroscopy (EIS)	32
2.7	Metal oxide nanostructures	33
	2.7.1 Zinc oxide nanoparticles	34
	2.7.1.1 Synthesis of zinc oxide nanostructures	34
	2.7.2 Titanium dioxide nanoparticles	35
	2.7.2.1 Crystal structure of titanium dioxide	36
	2.7.2.2 Synthesis of titanium dioxide nanostructures	38
2.8	Nanocomposites	39
2.9	Solvent dispersion and casting immobilization	39
2.10	Chitosan	40
2.11	Determination of carbohydrates	40
	2.11.1 Determination of glucose	41
2.12	Determination of amino acids	41
3	**Experimental procedures and materials for nanocomposites in electrochemical sensors**	**43**
3.1	Introduction	43
	3.1.1 Materials	43
3.2	Synthesis of nanocrystalline metal oxides by sol-gel method	44
3.3	Characterization of synthesized metal oxide particles	46
	3.3.1 Powder X-ray Diffraction (XRD)	46
	3.3.1.1 Measuring conditions	48
	3.3.2 Thermogravimetric Analysis (TGA)	48
	3.3.2.1 Measurement conditions	48
	3.3.3 Electron microscopy	49
	3.3.3.1 Scanning electron microscopy	50

	3.3.3.2	Transmission electron microscopy	50
	3.3.3.3	Measurement conditions	52
3.3.4	Fourier transform infrared spectroscopy		53
	3.3.4.1	Measurement conditions	55
3.3.5	UV–Visible spectroscopy		55
	3.3.5.1	Measurement conditions	55
3.4	Chemically Modified Electrodes (CMEs)		56
3.4.1	Pre-treatment of the electrodes		56
3.4.2	Preparation of phosphate buffer		57
3.4.3	Preparation of serum samples and real sample analysis		58
3.5	Fabrication of chemically modified electrodes		58
3.5.1	Fabrication of Co/ZnO/MWCNT/PCL/GCE electrode		58
3.5.2	Fabrication of TiO2-MWCNT/GCE electrode		59
3.5.3	Fabrication of Ni/PAN-MWCNT/GCE electrode		59
3.5.4	Fabrication of FC/CS-MWCNT/GCE electrode		59
3.6	Electrochemical measurements of modified electrodes		60
3.6.1	Electrochemical setup		60
3.7	Cyclic voltammetry		61

4 Electrocatalytic detection of glucose using Cobalt (II) ions immobilized ZnO/MWCNT/PCL nanocomposite film — **65**

4.1	Introduction		65
4.2	Experimental procedure		66
4.2.1	Characterization of ZnO nanoparticles		66
4.3	Results and discussion		68
4.3.1	Electrochemical measurement of Co/ZnO/MWCNT/PCL/GCE electrode		68
4.3.2	Electrochemical impedance spectroscopy		70
4.3.3	Electrocatalytic properties of Co/ZnO/MWCNT/PCL/GCE electrode		72
4.3.4	Interference study		74
4.3.5	Stability and reproducibility of Co/ZnO/MWCNT/PCL/GCE electrode		74
4.3.6	Determination of glucose in human blood serum		75
4.4	Chapter summary		75

5 Sensitive detection of L-Tryptophan using gelatin stabilized anatase titanium dioxide nanoparticles — **77**

5.1	Introduction		77
5.2	Experimental procedure		78
5.2.1	Synthesis of anatase TiO2 nanoparticles		78
5.2.2	Fabrication of TiO2-MWCNT/GCE		78
5.3	Results and discussion		79
5.3.1	Thermogravimetric analysis		79
5.3.2	X-ray diffraction to study TiO2 nanoparticles and poly dispersity index		80
5.3.3	Transmission of electron microscopy		81

		5.3.4	Fourier transform infrared spectroscopy (FTIR)	81
		5.3.5	UV-Vis absorption spectroscopy (UV-Vis)	82
	5.4		Electrocatalytic properties of TiO_2-MWCNT/GCE electrode	83
	5.5		Determination of *L*-Trp	83
	5.6		Chapter summary	85

6 Electrocatalytic detection of carbohydrates by using Ni(II)/PAN/MWCNT nanocomposite thin films **87**

	6.1		Introduction	87
	6.2		Experimental procedure	88
		6.2.1	Preparation of modified electrodes	88
	6.3		Results and discussion	88
		6.3.1	Properties of modified electrode	88
		6.3.2	Electrochemical Impedance Spectroscopy (EIS)	89
		6.3.3	The surface morphologic studies	91
		6.3.4	Electrocatalytic behavior of Ni/PAN-MWCNT/GCE nanocomposite	91
		6.3.5	Voltammetry response and calibration curve	93
		6.3.6	Interference study	94
		6.3.7	Reproducibility and stability of Ni/PAN-MWCNT/GCE electrode	95
		6.3.8	Determination of glucose and analytical recoveries of carbohydrates	95
	6.4		Chapter summary	95

7 Simultaneous detection of *D*-Penicillamine and *L*-Tryptophan using ferri/ferrocyanide doped chitosan-multi walled carbon nanotube nanocomposite **97**

	7.1		Introduction	97
	7.2		Experimental procedure	98
		7.2.1	Preparation of FC/CS-MWCNT/GC electrode	98
	7.3		Results and discussion	99
		7.3.1	$[Fe(CN)_6]^{3-/4-}$ redox probe	99
		7.3.2	Characterization of CS-MWCNT/GCE electrode	99
		7.3.3	Characterization of FC/CS-MWCNT/GC electrode in the presence of *D*-PA and *L*-Trp	103
		7.3.4	Effect of pH on *D*-PA and *L*-Trp oxidation	105
		7.3.5	Simultaneous detection of *D*-PA and *L*-Trp	105
		7.3.6	Stability and reproducibility of the FC/CS-MWCNT/GC electrode	105
		7.3.7	Real sample analysis	106
	7.4		Chapter summary	107

8 Outlook of nanocomposites in electrochemical sensors **109**

References 115

Preface

Nanotechnology has become one of the most important fields in science. The nanoparticles exhibit unique chemical, physical and electronic properties that are different from those of bulk materials, due to their small size and better architecture. Nanoparticles can be used to construct novel sensing devices; in particular electrochemical sensors. Electrochemical detection is highly attractive for the monitoring of glucose, cancer cells, cholesterol and infectious diseases. For the present investigation of current work, the nanomaterials of zinc oxide and anatase titanium dioxide were synthesized by simple and low-cost techniques using gelatin as an organic precursor via the sol-gel method. The analytical and spectroscopic techniques of thermogravimetric analysis, X-ray diffraction, transmission electron microscopy, scanning electron microscopy, Fourier transform infrared spectroscopy and ultraviolet–visible spectroscopy were used to characterize the architecture and properties of the synthesized nanomaterials. The result shows that, the metal oxide nanoparticles with high crystallinity can be obtained using this facile method. Gelatin plays a very important role in the formation of zinc oxide and anatase titanium dioxide nanoparticles.

Moreover, the composite film was prepared by using the synthesized zinc oxide, multi walled carbon nanotube and polycaprolactone on glassy carbon electrode (ZnO/MWCNT/PCL/GCE). The porous ZnO/MWCNT/PCL film was used as a supporting matrix to immobilize the Co(II) ions. The immobilized Co(II) ion exhibits an excellent electrocatalytic activity to glucose oxidation. The sensor responded linearly to glucose in the concentration of 5.0×10^{-5}–6.0×10^{-3} M with the detection limit of 1.6×10^{-5} M at 3σ using cyclic voltammetry. In addition, the composite film was prepared by synthesized TiO$_2$-NPs and multi walled carbon nanotube on glassy carbon electrode (TiO$_2$-MWCNT/GCE). The TiO$_2$-MWCNT/GCE responded linearly to L-Tryptophan in the concentration range of 1.0×10^{-6}–1.5×10^{-4} M with the detection limit of 5.2×10^{-7} M at 3σ using amperometric measurement.

A nickel (II) into porous polyacrylonitrile- multi walled carbon nanotubes composite modified glassy carbon electrode (Ni/PAN-MWCNT/GCE) was fabricated by simple drop-casting and immersing technique. Ni/PAN-MWCNT/GCE electrode showed good electrocatalytic activity toward the oxidation of carbohydrates (glucose, sucrose, fructose and sorbitol). The electrocatalytic response showed a wide linear range (10–1500 μM, 12–3200 μM, 7–3500 μM and 16–4200 μM for glucose, sucrose, fructose and sorbitol, respectively). At this juncture, I prepared a simple and efficient strategy for the fabrication of nanocrystalline composite films containing chitosan and multi walled carbon nanotube coated on a glassy carbon electrode.

The chitosan films are permeable to anionic $[Fe(CN)_6]^{3-/4-}$ redox couple. This modified electrode also showed an electrocatalytic effect on the simultaneous determination of D-penicillamine and L-Tryptophan. The detection limit of $0.9\,\mu M$ and $4.0\,\mu M$ for D-penicillamine and L-Tryptophan, respectively, makes this nanocomposite film very suitable for the determination of D-penicillamine and L-Tryptophan with good sensitivity and selectivity. In summary, the electrochemical sensors proposed in current work exhibited high sensitivity and selectivity for the continuous monitoring of analytes such as carbohydrates, D-penicillamine and L-Tryptophan. The unique nanocomposite-based films proposed in this book open new doors to the design and fabrication of high-performance electrochemical sensors.

List of figures

1.1 Synthetic methods of nanoparticles   4

1.2 Chemical structure of a) Polycaprolactone, b) Polyacrylonitrile, c) Polycaprolactone powder, d) Polyacrylonitrile powder   13

1.3 a) Chemical structure and b) powder of chitosan   14

1.4 a) Chemical structure and b) powder of gelatin   15

1.5 Carbohydrates: a) glucose, b) sucrose, c) fructose, d) sorbitol   15

1.6 Chemical structure of a) D-penicillamine, b) L-Tryptophan, c) D-penicillamine powder, d) L-Tryptophan powder   17

1.7 Typical cyclic voltammogram   19

1.8 Typical fitting Nyquist plot   21

1.9 Typical fitting according to Nyquist plot   21

1.10 Number of publications per year on application of nanomaterials in electrochemical sensors. Data extracted on 15th February 2013 through the Institute of Scientific Information (ISI) database   23

2.1 a) a single generic linear voltage sweep, b) a typical cyclic voltammogram   31

2.2 Chronoamperometric curve of TiO_2-MWCNT/GCE with successive addition of Tryptophan to a stirred 0.1 M PBS (pH 7.00). The inset is the calibration curve (Chapter 5)   32

2.3 Unit cell structures of the a) anatase and b) rutile crystals   37

3.1 Schematic diagram of sol-gel method for the synthesis of metal oxides   45

3.2 Work flow scheme for the synthesis of metal oxide nanoparticles by sol-gel method   45

3.3 X-Ray Diffractometer (Bruker D8 Advance)   47

3.4 Schematic diagram of thermal analysis (Mettler Toledo (SDTA-85 1e))   49

3.5 Scanning electron microscopy layout (FEI quanta 200F)   51

3.6 Schematic diagram of transmission electron microscopy (LEO-Libra 120)   52

3.7 Schematic diagram of Fourier transform infrared spectroscopy (Bruker, IFS 66v/s)   54

3.8 Schematic diagram of UV–Visible spectroscopy (Perkin Elmer, Lambda 35)   56

3.9 Fabrication of Co/ZnO/MWCNT/PCL/GCE electrode   58

3.10 Fabrication of TiO_2-MWCNT/GCE electrode   59

3.11 Fabrication of Ni/PAN-MWCNT/GCE electrode   60

3.12 Fabrication of FC/CS-MWCNT/GCE electrode 60

3.13 Three-electrode system a) reference electrode, b) counter electrode and c) working electrode (glassy carbon electrode) 61

3.14 Electrochemical Setup (Autolab, Metrohm) 62

3.15 Illustration of the concept of detection limit and quantitation limit by showing the theoretical normal distributions associated with blank, detection limit, and quantification limit level samples 63

4.1 XRD pattern of ZnO-NPs 67

4.2 TEM image of ZnO-NPs 67

4.3 FTIR spectra of the ZnO-NPs 68

4.4 UV-Vis absorbance spectrum of the ZnO-NPs 68

4.5 (A) Cyclic voltammograms of the (a) Co/PCL/GCE, (b) Co/ZnO/PCL/GCE, (c) Co/MWCNT/PCL/GCE and (d) Co/ZnO/MWCNT/PCL/GCE in 0.1 M NaOH solution at a scan rate of $50\,mV\,s^{-1}$. (B) Cyclic voltammograms of the (a) GCE, (b) PCL/GCE, (c) ZnO/PCL/GCE, (d) MWCNT/PCL/GCE, (e) ZnO/MWCNT/PCL/GCE 69

4.6 (A) Cyclic voltammograms of Co/ZnO/MWCNT/PCL/GCE in 0.1 M NaOH solution at different scan rates. The scan rates are: (a) 10, (b) 25, (c) 50, (d) 75, (e) 100, (f) 120 and (g) $150\,mV\,s^{-1}$, respectively. (B) The plot of cathodic and anodic peak currents vs scan rates 71

4.7 Impendence plots of (a) bare GCE, (b) PCL/GCE and (c) ZnO/PCL/GCE in the presence of 1.0 mM $[Fe(CN)_6]^{3-/4-}$ containing 0.1 M KCl as supporting electrolyte. Inset: (a) EIS of (a) MWCNT/PCL/GCE and (b) ZnO/MWCNT/PCL/GCE in the same condition 72

4.8 Cyclic voltammograms of Co/ZnO/MWCNT/PCL/GCE in the presence of (a) 0.05, (b) 0.08, (c) 0.20, (d) 0.40, (e) 0.80, (f) 3.00, (g) 4.00 and (h) 6.00 mM of glucose in 0.1 M NaOH solution at scan rate $50\,mV\,s^{-1}$. (B) Plot of anodic peak current vs. glucose concentration 73

5.1 TGA curves of TiO_2-NPs synthesized with (curve a) and without (curve b) gelatin from 50°C to 900°C 79

5.2 XRD patterns of TiO_2-NPs synthesized with (curve a) and without (curve b) gelatin 80

5.3 TEM images and the particle size of TiO_2-NPs synthesized with (curve a) and without (curve b) gelatin 81

5.4 FTIR spectra of TiO_2-NPs synthesized with (curve a) and without (curve b) gelatin 82

5.5 UV-Vis spectra of TiO_2-NPs synthesized with (curve a) and without (curve b) gelatin 83

5.6 Cyclic voltammograms of (a) GCE, (b) MWCNT/GCE, (c) TiO_2-MWCNT/GCE in 0.1 M PBS solution (pH 7.00) and 0.1 M KCl as supporting electrolyte at scan rate of $50\,mV\,s^{-1}$ 84

5.7 Current–time curve of TiO_2-MWCNT/GCE with successive addition of Trp to a stirred 0.1 M PBS (pH 7.00). The inset is the calibration curve 84

6.1 Cyclic voltammograms of the Ni/GCE (a), Ni/MWCNT/GCE (b), Ni/PAN/GCE (c) and Ni/PAN-MWCNT/GCE (d) in 0.1 M NaOH solution at a scan rate of $50\,mV\,s^{-1}$ 89

6.2   (A) Cyclic voltammograms of Ni/PAN-MWCNT/GCE in 0.1 M NaOH
       solution at different scan rates. The scan rates are: a) 10, b) 25, c) 50,
       d) 100, e) 200, f) 400, g) 600, h) 800 and i) 1000 mV s^{-1}, respectively.
       (B) The plot of cathodic and anodic peak currents vs. scan rates   90
6.3   Impedance plots of bare GCE (a), PAN/GCE (b) and
       PAN-MWCNT/GCE (c) in the presence of 1.0 mM [Fe(CN)$_6$]$^{3-/4-}$
       containing 0.1 M KCl as supporting electrolyte   91
6.4   SEM images of PAN/GCE (a), PAN-MWCNT/GCE (b) and
       Ni/PAN-MWCNT/GCE (c) electrodes   92
6.5   EDAX of Ni/PAN-MWCNT/GCE electrode   92
6.6   Cyclic voltammograms obtained for bare GCE in (a) absence and
       (b) presence of 1.0 mM of glucose (A), sucrose (B), fructose (C) and
       sorbitol (D). (c) as (b) and (d) as (b) at Ni/PAN-MWCNT/GCE in 0.1 M
       NaOH at scan rate 50 mV s^{-1} 93
6.7   (A) Cyclic voltammograms of Ni/PAN-MWCNT/GCE in the presence
       of a) 0.00, b) 0.01, c) 0.05, d) 0.20, e) 0.40, f) 0.70, g) 1.00, h) 1.30
       and i) 1.50 mM of glucose in 0.1 M NaOH solution at scan rate of
       50 mV s^{-1}. (B) Plot of anodic peak current vs. glucose concentration   94
7.1   Cyclic voltammograms of (A) 40 cycles of CS-MWCNT/GCE electrode
       in 1.0 mM [Fe(CN)$_6$]$^{3-/4-}$ and 0.1 M KCl at 0.05 V/s and (B) 20 cycles
       of CS-MWCNT/GCE electrode post [Fe(CN)$_6$]$^{3-/4-}$ exposure in
       0.1 M KCl   100
7.2   The cyclic voltammogram of (a) GCE, (b) CS/GCE, (c) MWCNT/GCE
       and (d) CS-MWCNT/GCE electrodes in 2.0 mM [Fe(CN)$_6$]$^{3-/4-}$ and
       0.1 M KCl as supporting electrolyte at a scan rate 0.05 V/s   101
7.3   Nyquist plots recorded at (a) bare GCE, (b) CS/GCE and
       (c) CS-MWCNT/GCE electrode in the presence of 1.0 mM [Fe(CN)$_6$]$^{3-/4-}$
       solution containing 0.1 M KCl as supporting electrolyte   102
7.4   SEM images of (a) GCE, (b) CS/GCE, (c) CS-MWCNT/GCE and (d)
       FC/CS-MWCNT/GCE electrodes   102
7.5   UV-Vis spectra of (a) chitosan in 0.1 M acetic acid solution mixture and
       (b) mixture acidic solution of chitosan and multi walled carbon
       nanotube   103
7.6   Cyclic voltammograms of (A) pH 7.00 PBS in (a) absence and
       (b) presence of 0.1 mM of D-PA at GCE and (c) as (a) and (d) as
       (b) at FC/CS-MWCNT/GCE electrode at scan rate of 0.02 V/s; (B) pH
       7.00 PBS in (a) absence and (b) presence of 0.1 mM of L-Trp at GCE,
       (c) in presence of 0.3 mM D-PA + 0.3 mM L-Trp mixture at
       CS-MWNT/GCE, (d) as (a) and (e) as (b) at FC/CS-MWCNT/GC
       electrode   104
7.7   DPVs of (A) 100 μM Trp in presence of different concentrations of
       D-PA: (a) 0.000, (b) 0.003, (c) 0.007, (d) 0.020, (e) 0.040, (f) 0.080,
       (g) 0.160 and (h) 0.300 mM and (B) 3.0 μM D-PA in presence of
       different concentrations of Trp: (a) 0.007, (b) 0.028, (c) 0.060,
       (d) 0.120, (e) 0.180, (f) 0.220 and (g) 0.300 mM at the FC/CS-
       MWCNT/GC electrode in pH 7.00 PBS. (A$'$) and (B$'$) are the Plot of
       anodic peak currents vs. D-PA and Trp concentrations, respectively   106

List of tables

1.1   Roles of nanoparticles in electrochemical sensor systems   6
1.2   Synthesis methods of carbon nanotubes   10
2.1   Crystallographic parameters of ZnO (ICDD card number: 01-079-0206)   34
2.2   Synthesis methods of ZnO nanoparticles   35
2.3   Crystallographic parameters of anatase (ICDD card number: 01-073-1764)   36
2.4   Crystallographic parameters of rutile (ICDD card number: 01-076-0324)   36
3.1   List of chemical compounds used   44
4.1   Anodic and cathodic potential, the half-wave potential and the peak-to-peak separation potential for redox couple in cyclic voltammogram of Figure 4.5A   70
4.2   The analytical recoveries of glucose solutions added to 0.1 M NaOH solution, suggesting better accuracy of the method   74
4.3   Assay of glucose in human blood serum samples and recovery of glucose in 0.1 M NaOH solution spiked with different concentrations   75
6.1   The Anodic and cathodic potential, the half-wave potential and the peak-to-peak separation potential for redox couple of cyclic voltammogram of Figure 6.1   90
6.2   Analytical parameters for voltammetric determination of carbohydrates at different modified electrodes   95
6.3   Assay of glucose in human blood serum samples and recovery of carbohydrates in 0.1 M NaOH solution spiked with different concentrations   96
7.1   The Anodic and cathodic potential, the half-wave potential and the peak-to-peak separation potential for redox couple of cyclic voltammogram of Figure 7.2   103
7.2   Analytical parameters for voltammetric determination of D-PA at different modified electrode   107
7.3   Measurement results of D-PA in commercial tablet (Dr. Abidi pharmaceutical laboratories Co.) and recovery of D-PA and Trp in 0.1 M phosphate buffer solution spiked with different concentrations   107
8.1   Summary of the electrochemical sensors properties which were fabricated in this research   112

List of equations

1.1 $E_{pa} - E_{pc} = \dfrac{59}{n} V$ 19

1.2 $E = E^\circ + \left\{ \dfrac{59}{n} \right\} \log\left(\dfrac{a_B}{a_{B^{n-}}} \right)$ 19

1.3 $I_p = \dfrac{dQ}{dt} = \dfrac{dN}{dt} nF$ 20

1.4 $N = \dfrac{Q}{nF}$ 20

1.5 $i = nFk \exp\left(\pm \alpha nF \dfrac{\Delta V}{RT} \right)$ 21

1.6 $i = i_o \cdot \left\{ \exp\left[\dfrac{\alpha_a nF\eta}{RT} \right] - \exp\left[-\dfrac{\alpha_c nF\eta}{RT} \right] \right\}$ 22

1.7 $i = \dfrac{nFD}{\delta} C^*$ 22

2.1 $I = nFAc_0 \sqrt{\dfrac{D}{\pi t}}$ 32

2.2 $Z(jw) = \dfrac{U(jw)}{I(jw)} = Z_r(w) + jZ_i(w); \quad w = 2\pi f$ 33

3.1 $D = \dfrac{K\lambda}{\beta_{hkl} \cos\theta}$ 47

3.2 $\beta_{hkl} = (B_{hkl} - b)$ 47

4.1 $(E) = \dfrac{hc}{\lambda}$ 67

6.1 $\Gamma = Q/nFA$ 89

List of symbols and abbreviations

FESEM	Field emission scanning electron microscopy
FTIR	Fourier transform infrared spectroscopy
FWHM	Full width at half maximum
NP	Nanoparticle
TGA	Thermogravimetric analysis
TEM	Transmission electron microscopy
HRTEM	High resolution transmission electron microscopy
UV-Vis	Ultraviolet–Visible spectroscopy
XRD	X-ray diffraction
EDAX	Energy dispersive X-ray
CVD	Chemical vapour deposition
ED	Electrodeposition
DMF	Dimethyl formamide
PCL	Polycaprolactone
PAN	Polyacrylonitrile
PEG	Polyethylene glycol
CTAB	Cetyltrimethyl ammonium bromide
PSS	Poly (sodium 4-styrene-sulfonate)
L-Trp	L-Tryptophan
D-PA	D-Penicillamine
CV	Cyclic voltammetry
EIS	Electrochemical impedance spectroscopy
MWCNT	Multi walled carbon nanotubes
SWCNT	Single walled carbon nanotubes
CME	Chemically modified electrodes
DPV	Differential pulse voltammetry
ZnO	Zinc oxide
GCE	Glassy carbon electrode
WE	Working electrode
RE	Reference electrode
CE	Counter electrode
DC	Direct current
AC	Alternating current
GPES	General purpose electrochemical system
FRA	Frequency response analysis

RSD	Relative standard deviation
LOD	Limit of detection
LDR	Linear dynamic range
PDI	Poly dispersity index
PBS	Phosphate buffered saline
L-Trp	L-Tryptophan
(α)	Electron transfer coefficient
E^o	Formal peak potential
ΔE_p	Peak-to-peak separation potential
E_{pc}	Cathodic peak potential
E_{pa}	Anodic peak potential
I_{pc}	Cathodic peak current
I_{pa}	Anodic peak current
$E_{1/2}$	Half peak potential
F	Faraday's constant
w	Angular frequency
$Z_r(w)$	Real impedance component
$Z_i(w)$	Imaginary impedance component
E	Potential
R	Resistance
I	Current
Γ	Surface coverage
Q	Charge
R_{ct}	Electron transfer resistance
R_s	Solution resistance
C_{dl}	Double-layer capacitance
W	Warburg diffusion
η	Activation overpotential
δ	Diffusion layer thickness

Chapter 1

Introduction of nanocomposites in electrochemical sensors

1.1 INTRODUCTION TO NANOTECHNOLOGY

Nanotechnology originates from a Greek word meaning 'dwarf'. Nanometer is one billionth (10^{-9}) of a metre, which is very tiny. It is only the length of a few hydrogen atoms or about one hundred thousandth of a human hair. Nanotechnology involves research and development at the atomic and macromolecular levels in the length scale of approximately 1 to 100 nm. The U.S. National Nanotechnology Initiative (NNI) definition decrees that anything smaller than 100 nanometers with novel properties is termed as Nanotechnology. Nanotechnology exploits the numerous progresses in chemistry, physics, materials science and bioscience to create novel materials that have unique properties due to their size, structure and architectures and are determined on the nanometer scale. Some of these materials have already found their way into industrial products.

The term nanotechnology was first coined by N. Taniguchi in the year 1974 (Taniguchi, 1974). The first report about the creation of these new materials was published in 1985 (Kroto *et al.*, 1985). Various aspects of synthesis, study and application of nanomaterials in different fields mean that even in such a short period of time it has been possible to take a considerable step forward and gain impressive results (Shameli *et al.*, 2012).

Nanotechnology is referred to as a wide-ranging technology because of its advanced impact in almost all industries. It will offer superior materials with long-term duration, safety and smaller products for home, medicine, communication, agriculture and transportation. Realization of bottom-up approach (chemical approach) of molecular nanotechnology will further attract the interest of material scientists working in the field of nanotechnology (Chekin *et al.*, 2013a).

According to the National Nanotechnology Initiative (2011), the estimated amount is found in all over the world is around $65 billion, due to the vast opportunities in nanotechnology, which offers to create the new features and functions. Recently, the material scientists have classified nanotechnology in three directions: In physics, the field of microelectronics is moving towards smaller sizes and is already at submicron line widths; processors in computing systems will need nanometer line widths in future as miniaturization proceeds (Ramimoghadam *et al.*, 2014b).

In chemistry, improved knowledge of complex systems had led to new catalyst, membrane sensor and coating technologies which depend on the ability to tailor the structures at atomic and molecular levels (Julkapli and Bagheri, 2015). In biology, living systems have sub-units with sizes between micron and nanometer scales and these can be combined with non-living nanostructured materials to create new devices (Shameli *et al.*, 2015).

In recent times, nanotechnology has become one of the most exciting fields in physics, chemistry and material science (Muhd Julkapli *et al.*, 2014). A wide variety of nanomaterials especially nanoparticles with different properties have found a wide application in many kinds of chemical and electrochemical methods. Nanoparticles exhibits unique physical-chemical and electronic properties that are different from bulk materials due to their small size and unique structure, and also can be used to construct novel materials to improve the sensing devices like electrochemical and biosensors (Bagheri *et al.*, 2015b; Bagheri *et al.*, 2012b; Chekin *et al.*, 2012b). Nowadays, a large number of nanoparticles with different sizes and compositions are available, which facilitates their applications in electrocatalysis and electrolysis (Chekin *et al.*, 2012a). Different kinds of nanoparticles and sometimes the same kind of nanoparticles can play a different role in different electrochemical sensing systems such as enzyme sensors, DNA sensors and immunosensors. Generally, metal nanoparticles have excellent catalytic properties and conductivity which make them more suitable for acting as electronic wires to improve the electron transfer between redox centres in proteins and electrode surfaces, and also as catalysts to increase the electrochemical reactions. Oxide nanoparticles are frequently used to immobilize biomolecules due to their biocompatibility, while semiconductor nanoparticles are regularly used as tracers for electrochemical analysis.

1.2 APPLICATION OF NANOTECHNOLOGY IN CATALYSIS

Nanotechnology has gained substantial popularity recently due to the rapidly developing techniques both to synthesize and characterize materials and devices at the nano-scale, as well as the promises that such technology offers to substantially expand the achievable limits in many different fields including medicine, electronics, chemistry, and engineering. In the literature, there are constantly reports of new discoveries of unusual phenomena due to the small scale and new applications (Termeh Yousefi *et al.*, 2014). Nano-size noble metal particles have occupied a central place in catalysis for many years, long before recognition of nanotechnology. Thus, it is fitting to critically evaluate the impact of such development on catalysis (Bagheri *et al.*, 2013b). One of the most important properties of a catalyst is its active surface where the reaction takes place (Termehyousefi *et al.*, 2015b). When the size of the catalyst is decreased, the active surface increases; the smaller the catalyst particles, the greater the ratio of surface to volume (Amiri *et al.*, 2014). The higher the catalyst's active surface, the greater is the reaction efficiency. Research has illustrated that the spatial organization of active sites in a catalyst is very important. Molecular structure and nanoparticle size, can be controlled using nanotechnology (Bagheri *et al.*, 2013a). Achieving a high degree of selectivity in catalysis is also recognized as a critical challenge for the future (Zhou *et al.*, 2003). Transition metal oxide nanoparticles are good catalysts in redox reactions due to their ability to readily move from one oxidation state to another (Gholamrezaei *et al.*, 2014).

A major goal in catalysis research is to design catalysts that can achieve perfect selectivity and desirable activity (Ramimoghadam et al., 2014a). Between activity and selectivity, it is commonly accepted that the latter is much more difficult to achieve and control. Thus, it is the focus of our discussion. A reaction of perfect selectivity would generate no waste products, thereby reduce energy and process requirements for separation and purification (Luo et al., 2006).

1.3 NANOMATERIALS

In the last two decades, there has been enormous progress in the synthesis of nanoparticles research. Nanoparticles are defined as a group of atoms having size ≤ 100 nm (Daniel & Astruc, 2003; Siegel & Fougere, 1995). Nanoparticle size ranges between atoms and molecules and bulk. The chemical (Cox et al., 1988), physical (Nieh & Wadsworth, 1990; Philip, 2001; Schmidt Martin, 1998; Siegel & Fougere, 1995), mechanical, thermal, electronic and magnetic properties of nanometer sized particles are very different from their bulk counterparts and can be tuned by changing the size and shape. Owing to these shape and size dependent properties, nanoparticles are very useful for many applications such as catalysis, optoelectrics, chemosensors and single transistors (Alivisatos, 1996; Leutwyler et al., 1996). Two major effects that are responsible for the shape and size dependent properties of nanoparticles are increase in surface to volume ratio of nanoparticles and quantum size effects which influence their electronic structure (Alivisatos, 1996). The size at which the nanoparticle behaves like its bulk depends on the type of materials. In metals, a few tens of atoms are adequate to make the nanoparticles behave like bulk but in semiconductors this number is much larger (Alivisatos, 1996). For metallic gold, if the size of bulk materials is transformed into particles of size <1.6 nm (Wolfe et al., 2007), a transition from metal to insulator takes place. The reason for this transition is the relocation of valence free electrons from continuous energy states to discrete energy states (Chekin et al., 2012c). Hence, the size dependent optical and electronic properties of nanoparticles are due to an increase discretization of energy levels with decrease in size (Azizian et al., 2012).

Many kinds of nanoparticles including metal nanoparticles, oxide nanoparticles, semiconductor nanoparticles and even composite nanoparticles have been widely used in electrochemical sensors and biosensors (TermehYousefi et al., 2015d). Depends on the role of these nanoparticles play in different electrochemical sensing systems based on their unique properties, the basic functions of nanoparticles can be mainly classified as immobilization of biomolecules, catalysis of electrochemical reactions, enhancement of electron transfer, labelling biomolecules and acting as reactant (Chekin et al., 2012d).

1.4 SYNTHESIS OF NANOMATERIALS

Synthesis and assembly strategies of nanomaterials mostly accommodate precursors from liquid, solid or gas phase, employ chemical or physical deposition approaches, and similarly rely on either chemical reactivity or physical compaction to integrate the nanostructure building blocks within the final material structure (Bagheri et al.,

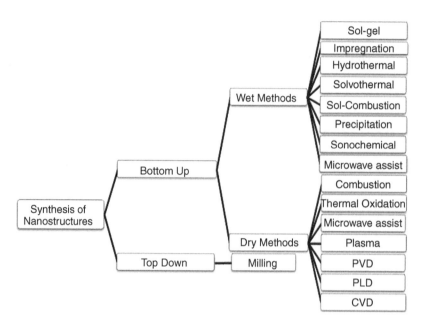

Figure 1.1   Synthetic methods of nanoparticles.

2014d). The bottom-up approach of nanomaterials synthesis first forms the nanostructured building blocks (nanoparticles) and then assembles these into the final material (Ramimoghadam *et al.*, 2015). An example of this approach is the formation of powder components through aerosol and sol-gel techniques and then the compaction of the components into the final material (Amiri *et al.*, 2015). Nanoparticles with diameters ranging from 1 nm to 10 nm with consistent crystal structure, surface derivatisation, and a high degree of monodispersity can be processed by these techniques (Bagheri *et al.*, 2014c).

Synthetic methods of nanoparticles can be classified into three general categories; they are wet, dry and milling process as shown in Figure 1.1. In wet and dry synthesis, the nanoparticles are generally produced through a bottom up approach, whereas in the milling process, the nanoparticles are produced via top down approach by mechanically breaking down the larger particles (Rotello, 2004). In all the methods, the size distribution of nanoparticles is very important and can be modified by adjusting the process parameters. Milling is very energy-intensive and it may not be suitable for some materials such as pure metals because they are malleable. In the precipitation method, it is necessary to add capping ligands to the solution for controlling the growth of nanoparticles. These ligands bind to the surface of particles and must be removed in a separate processing step. One of the biggest problems during the high temperature heating process is an agglomeration of nanoparticles.

In this research work, the sol-gel method was applied for the preparation of nanostructured materials using gelatin as the polymerizing agent.

1.5 NANOPARTICLES

Nanoparticles are particles between 1 and 100 nanometers in size. In nanotechnology, a particle is defined as a small object that behaves as a whole unit with respect to its transport and properties (Bagheri *et al.*, 2014a). Particles are further classified according to diameter. Ultrafine particles are the same as nanoparticles and between 1 and 100 nanometers in size, fine particles are sized between 100 and 2,500 nanometers, and coarse particles cover a range between 2,500 and 10,000 nanometers. Nanoparticle research is currently an area of intense scientific interest due to a wide variety of potential applications in the biomedical, optical and electronic field (Abdolvahabi *et al.*, 2012a).

Nanoparticles are of great scientific interest as they are, in effect, a bridge between bulk materials and atomic or molecular structures. A bulk material should have constant physical properties regardless of its size, but at the nano-scale size-dependent properties are often observed (Termeh Yousefi *et al.*, 2014). Thus, the properties of materials change as their size approaches the nanoscale and as the percentage of atoms at the surface of a material becomes significant. For bulk materials larger than one micrometer, the percentage of atoms at the surface is insignificant in relation to the number of atoms in the bulk of the material. The interesting and sometimes unexpected properties of nanoparticles are therefore largely due to the large surface area of the material, which dominates the contributions made by the small bulk of the material (Upadhyay *et al.*, 2014).

The unique physical and chemical properties of nanoparticles make them extremely sufficient to improve and design sensing devices and in especially electrochemical sensors and biosensors. The nanoparticles such as semiconductors, metal and metal oxide nanoparticles have been used for designing electrochemical sensors and biosensors. The main reason for using nanoparticles is its ability to immobilize biomolecules, catalysis of electrochemical reactions, enhancement of electron transfer between an electrode surface and proteins, labeling of biomolecules, and acting as a reactant.

Many kinds of nanoparticles including metal, metal oxide, semiconductor nanoparticles and even composite nanoparticles have been extensively used in electrochemical sensors and biosensors. These nanoparticles play different roles in different electrochemical sensing systems based on their unique properties. The basic functions of nanoparticles can be mainly classified as catalysis of electrochemical reactions, enhancement of electron transfer, immobilization of biomolecules, labeling biomolecules and acting as reactant. The role of different nanoparticles has played in electrochemical sensor systems are summarized in Table 1.1.

1.6 SYNTHESIS OF NANOPARTICLES

Nanoparticles have garnered a great deal of attention from the perspective of both basic and developmental science in a vast range of fields. This is due to the fact these nanoparticles exhibit size-dependent electrical, optical, magnetic and catalytic phenomena that cannot be realized by their bulk counterparts. Iron oxide nanoparticles, as an example, exhibit super-paramagnetism at room temperature while semiconductor "quantum dots" exhibit the quantum confinement effect (Chekin *et al.*, 2014c).

Table 1.1 Roles of nanoparticles in electrochemical sensor systems.

Functions	Properties used	Typical NPs	Sensor advantages	Typical examples	Reference
Biomolecules, immobilization	Biocompatibility; Large surface area	Metal NPs (Au, Ag); Oxide NPs (SiO$_2$, TiO$_2$)	Improved stability	Antibody immobilized onto Au NPs	(Zhuo et al., 2006)
Catalysis of reaction enhancement of electron transfer	High surface energy conductivity; tiny dimensions	Metal NPs (Au, Ag); Oxide NPs (ZrO$_2$, TiO$_2$)	Improved sensitivity & selectivity; direct electrochemistry of proteins	H$_2$O$_2$ sensor based on Prussian blue NPs with a sensitivity of 103.5 μA mM^{-1} cm^{-2}	(Fiorito et al., 2004)
Labeling biomolecules	Small size; modifiability	Semiconductor NPs (CdS, PbS); metal NPs (Au, Ag)	Improved sensitivity; Indirect detection	DNA sensor labeled with Ag NPs	(Cai et al., 2002)
Acting as reactant	Chemical activity	Oxide NPs (MnO$_2$)	New response mechanism	Lactate biosensor with MnO$_2$ NPs	(Zhao et al., 2005)

For spherical particles, the surface area/volume ratio is inversely proportional to the radius so a substantial reduction in particle size leads to a dramatic increase in surface area.

Nanoparticles size ranging from 1 to 100 nm has been actively synthesized and studied over the last 10 years. It has more scientific interest because of size dependent properties such as quantum confinement in semiconductor nanoparticles and surface plasmon resonance in some metal nanoparticles. The synthesis of monodisperse nanoparticles is of key importance since the properties of nanoparticles are dependent on their size (Wang *et al.*, 2005). There are two approaches for synthesis of nanoparticles: top-down (physical method), and bottom-up (chemical method). The top-down approach includes ball milling. However, the nanoparticles synthesized by physical method have a relatively broad size distribution and various particle shapes. In contrast, the bottom-up approaches can be used to synthesize uniform nanoparticles (Wang *et al.*, 2005).

1.6.1 Sol-gel method

The sol-gel process is a wet-chemical technique widely used in the fields of ceramic engineering and materials science. This method is used primarily for the fabrication of materials (typically metal oxides) starting from a colloidal solution (sol) that acts as the precursor for an integrated network (or gel) of either discrete particles or matrix polymers. Common precursors are metal alkoxides and metal salts (such as chlorides, nitrates and acetates), which undergo various forms of hydrolysis and polycondensation reactions.

In the sol-gel procedure, the solution (or sol) gradually evolves towards the formation of a gel-like system containing both a solid phase and liquid phase whose morphologies range from discrete particles to continuous polymer matrixes. In the case of the colloid, the particle density may be so low where a significant amount of fluid may need to be removed primarily from the gel-like properties to be recognized. Removal of the remaining liquid phase (solvent) requires a drying process, which is typically accompanied by a significant amount of densification and shrinkage. The rate at which the solvent can be removed is eventually determined by the distribution of porosity in the gel. The final microstructure of the final component will clearly be strongly influenced by changes imposed upon the structural template during this phase of processing.

Afterwards, a thermal treatment is often necessary in order to favor further polycondensation, structural stability and enhanced mechanical properties by final sintering, densification and grain growth. One of the distinct advantages of using this methodology as opposed to the more traditional processing techniques is that densification is often achieved at a much lower temperature.

The precursor sol can be either deposited on a substrate to form a film (by dip coating or spin coating), cast into a suitable container with the desired shape (to obtain monolithic ceramics, membranes, aerogels, glasses, fibers), or directly used to synthesize powders such as microspheres, nanospheres. The sol-gel approach is a cheap and low-temperature method that allows production of fine chemical composition product. Sol-gel method can be used in ceramics manufacturing as an investment casting material, or for producing very thin films of metal oxides for different purposes. Sol-gel synthesized materials have various applications in electronics, energy, biosensors, optics, and medicine such as controlled drug release, reactive material and separation technology such as chromatography.

The sol-gel method is a wet-chemical technique extensively employed recently in the fields of ceramic engineering and materials science. Such methods are utilized mainly in the fabrication of materials such as metal oxides which starting from a chemical solution that acts as the precursor for a gel of either discrete particles or matrix polymers.

Typical precursors are metal chlorides and metal alkoxides, which undergo different forms of hydrolysis and polycondensation reactions. The formation of a metal oxide involves connecting the metal centres with either oxo (M–O–M) or hydroxo (M–OH–M) bridges, and generating metal-oxo or metal-hydroxo polymers in the solution. Hence, the sol evolves towards the formation of a gel-like diphasic system containing both liquid and solid phases whose morphologies range from discrete particles to continuous polymer matrixes. For colloid, the particle density may be so low that a considerable amount of fluid may need to be removed initially for the gel-like properties to be recognized.

This can be achieved through various ways. The simplest method is to allow time for sedimentation to happen, and then pour off the remaining liquid. Another good way is centrifuging which can also be used to accelerate the process of phase separation.

Removal of solvent phase requires a drying process, which is typically accompanied by a considerable amount of solvent decrease and densification. The rate of removal of solvent phase can ultimately determine the distribution of porosity in the gel. The final microstructure of the ultimate component will be strongly influenced by changes imposed upon the structural template during this phase of processing. After that, a thermal treatment is necessary for further poly-condensation and to improve the structural stability and mechanical properties by final sintering, densification, and grain growth. One of the advantages of this methodology is that densification is often achieved at much lower temperatures.

The precursor sol can be either deposited on a substrate to form a film by methods of dip coating or spin coating, cast into a suitable container with the desired shape to obtain monolithic ceramics, membranes, glasses and aerogels or could be used to synthesize powders such as microspheres and nanospheres. The sol-gel method is a cheap and low-temperature technique which controls the chemical composition of the products finely.

The intrinsic advantages of the sol-gel method are better homogeneity from raw materials, better purity of final product, lower temperature of preparation, good mixing for multi-component systems, effective control of particle size, shape, and properties, as well as the possibility of designing the material structure and property through proper selection of sol-gel precursors.

1.7 APPLICATION OF NANOPARTICLES IN ELECTROCHEMICAL SENSORS

Many nanoparticles, especially metal nanoparticles, have excellent catalytic properties combining, as they do, the intrinsic catalytic properties of the metal with the nanoparticles properties of high surface area/volume ratio (Abdolvahabi *et al.*, 2012b). The introduction of nanoparticles with catalytic properties into electrochemical sensors and biosensors can decrease overpotentials of many analytically important electrochemical reactions, and even realize the reversibility of some redox reactions, which are irreversible at common unmodified electrodes (Bagheri *et al.*, 2015a). Some of the non-metal nanoparticles that have special catalytic properties can also be applied in electrochemical analysis systems. For example, a carbon paste electrode doped with copper oxide nanoparticles was developed for the detection of amikacin based on the catalytic properties of the copper oxide nanoparticles, and the oxidation current of amikacin at the prepared electrode was about 40 times higher than that at a bulk copper oxide modified carbon paste electrode (Ramimoghadam, 2015).

The exclusive physical and chemical properties of nanoparticles make them extremely suitable for designing new and improved sensing devices, especially in electrochemical sensors and biosensors. Several kinds of nanoparticles such as metal, metal oxide and semiconductor nanoparticles have been used for constructing the electrochemical and biosensors and also these nanoparticles play various roles in different sensing systems. The important functions provided by nanoparticles including immobilization of biomolecules, catalysis of electrochemical reactions and the improvement of electron transfer of electrode surfaces, labelling of biomolecules and even acting as reactant.

1.8 CHEMICALLY MODIFIED ELECTRODES (CMES)

Modification of the surface of electrodes is necessary in order to impose desired new properties (Murray, 1980). This approach is very useful in basic electrochemical studies as well as in storage, electro-synthesis, corrosion protection, molecular electronics and electrochromic displays.

In the present studies, chemically modified electrodes were designed to have electrocatalytic properties which enhance the detection signal and lower the overpotential necessary for the detection. The electrocatalysis at modified electrode accelerates the rate of electron transfer, which is slow when using the same potential at the unmodified electrode. The electrocatalytic modified electrodes are prepared by the immobilization of a catalyst on the electrode surface. The electrocatalysis at such electrodes is accomplished by charge mediation, which requires that the difference in the formal potentials of the redox mediator and analyte be thermodynamically favorable. For example, in the case of electro-oxidation processes, the formal potential of the redox mediator should be more positive than the formal potential of the analyte. The rate of electrocatalysis depends on the difference in their respective formal potentials. The reaction between the electrode-confined redox mediator and an analyte results in a catalytic current, which shows a chemical amplification of the detection signal. The amperometric measurements are made on a constant applied potential with the analyte being transported to the electrode surface by convection or diffusion. Such measurements yield currents that are directly proportional to the analyte concentration in a solution. This proportionality provides the analytical basis for the quantification of analytes. A variety of electrocatalytic electrodes have been prepared by either covalently attaching mediators (small redox active molecules) directly to the electrode surface or by immobilizing them in thin polymeric films cast on the surface.

In addition to increasing current, electrocatalytic electrodes decrease the potential necessary for the detection by changing the reaction mechanism. This is particularly important for kinetically slow-moving electrode reactions of analytes that require a high overpotential (a potential exceeding the analyte's formal potential). Lowering the potential allows the reaction to proceed at adequately high rates that produce an analytically useful amount of current. The acceleration of such reactions by attaching redox mediators allows the detection of these analytes at low extreme potentials because the efficiently catalyzed reactions typically occur close to the formal potential of the redox mediator. The less extreme potentials are advantageous because they can improve the selectivity and decrease electrode fouling due to radical polymerization. The latter frequently takes place in the case of un-mediated reactions requiring more extreme potentials at bare electrodes.

1.9 CARBON NANOTUBES

Carbon nanotubes are relatively new nanomaterials that display attractive structural, mechanical, and electronic properties, including improved electrochemical activity. They are allotropes of carbon which is made up of graphene sheets rolled up to form multi-walled (2–100 nm diameters) or single-walled (0.4–1.5 nm diameter) cylindrical structures with a characteristic length in the order of several microns.

Table 1.2 Synthesis methods of carbon nanotubes.

Methods	Arc discharge	Laser ablation	CVD
Founder	Iijima (1991)	Guo et al. (1995)	Yacaman et al. (1993)
Way of growth	CNT growth on graphite electrodes during direct current arc discharge evaporation of carbon in presence of an inert gas	Vaporization of a mixture of carbon (graphite) and transition metals located on a target to form CNTs	Fixed bed method: acetylene decomposition over graphite-supported iron particles at 700°C
Advantages	Simple, inexpensive	Relatively high purity CNTs, room temperature synthesis option with continuous laser	Simple, inexpensive, low temperature, high purity and high yields, aligned growth are possible, fluidized bed technique for large scale
Disadvantages	Purification of crude product is required, the method cannot be scaled up, must have high temperature	The method is only adapted to lab-scale. Crude product purification required	CNTs are usually MWCNTs, parameters must be closely watched to obtain SWCNTs

Carbon nanotubes can be produced by three main methods known as electric arc discharge, laser ablation and chemical vapor deposition. Electric arc discharge and laser ablation depend on the condensation of carbon atoms generated from evaporation of solid carbon sources such as graphite, at very high temperatures and require no metal catalyst.

They produce the highly entangled bundles of nanotubes that are difficult to purify, although the nanotubes have very few defects. The chemical vapor deposition method depends on thermal decomposition of a liquid-phase or gas-phase carbon source such as ethylene on a catalytic surface like Fe, Co and Ni nanoparticles. The carbon nanotubes synthesized by chemical vapor deposition are easier to purify, but have more defects than the carbon nanotubes formed by an arc discharge. Table 1.2 summarises various synthesis methods of carbon nanotubes together with advantages and disadvantages for each method.

The carbon-carbon bond in carbon nanotubes are essentially sp^2 hybridized, which is a circular curvature and concentric tubes causes σ-π transition (rehybridization) that is three σ bonds are slightly out of plane. To compensate for this slight bending, the p orbitals are more delocalized outside the tube leading to a rich π-electron conjugation. As the nanotube diameter decreases this compensating effect increases. The σ-π rehybridization and the presence of structural surface defects including atomic vacancies such as pentagon and n-heptagon pairs and their edges induce the local perturbations in the electronic structure and thus influence the chemical properties of carbon nanotubes. For example, the oxidative acid treatment of nanotubes opens their capped ends and results in the formation of various carbon-oxygen functional groups on their surface.

1.9.1 Electrochemical properties of carbon nanotubes

The electrochemical qualities of carbon based electrodes are considerably reliant on the surface properties as the creation of specific surface functional groups (especially oxygen-containing groups) can significantly increase the rate of electron transfer. Based on their specific structures, two distinct surface regions exist in carbon nanotubes, the side walls and the end cap.

In view of the fact that carbon nanotubes can be seen as graphene sheets rolled into tubes, the electrochemical properties of carbon nanotubes are comparable to the basal planes of pyrolytic graphite. For carbon nanotubes, the defect free structure makes the whole tubes possess almost the same properties as that of the basal plane of pyrolytic graphite except that the cap regions may be more reactive due to the much higher curve strain than the sidewall. The opening of the ends by physical/chemical treatments on carbon nanotubes produces a diversity of oxygen-containing groups, which possess the properties similar to the edge places of basal planes of pyrolytic graphite.

The electrochemical properties of carbon nanotubes are determined by their electronic structure including the density of electronic states and structure of their surface. These properties affect the kinetics and thermodynamics of electron transfer reactions in carbon nanotubes. There have been many reports about the remarkable electroanalytical benefits of electrodes modified with carbon nanotubes including the increased sensitivity, the lower detection limit, decrease over potentials and resistance to surface poisoning. However, there is a degree of discussion in previous literature as to whether such behavior is due to the new special properties of carbon nanotubes or a reflection of the presence of known and reactive graphitic edge plane in carbon nanotubes.

As an example, carbon nanotubes have been broadly employed in constructing diverse electrochemical sensors. Compared with the conventional scale materials and other types of nanomaterials, the special nanostructural properties lend carbon nanotubes some overwhelming advantages in fabricating electrochemical sensors. They include large specific area producing high sensitivity, tubular nanostructure and chemical stability allowing the fabrication of ultrasensitive sensors consisting of only one nanotube. It has good biocompatibility that is suitable for constructing electrochemical biosensors, especially for facilitating the electron transfer of redox proteins and enzymes. Modifiable ends and sidewalls of CNT provide a chance for fabricating multi-functioned electrochemical sensors via the construction of functional nanostructures, and the possibility of achieving miniaturization and the possibility of constructing ultrasensitive nanoarrays.

1.9.2 Purification of carbon nanotubes

Freshly prepared carbon nanotubes have impurities ranging from metal catalysts to fullerenes and amorphous carbon. In order to employ the carbon nanotubes for any chemical or biochemical purpose, there must be further purification. The multi walled carbon nanotubes are promising material for many technological applications considering their unique structure and possibility of modifications influencing their physical and chemical properties.

Due to their large specific surface area and conductivity, multi walled carbon nanotubes can be successively applied as a support for metallic catalysts in electrooxidation reaction in fuel cells and other variety of reactions in gas storage.

The modifications of multi walled carbon nanotubes material deal with several problems. First, the as-prepared synthesized material requires purification, that is removal of amorphous carbon, traces of catalysts and catalyst supports. Such procedures are based on chemical and electrochemical reactions which result in obtaining a material containing carbon defects that is sp^3-hybridized carbon atoms and other carbon, oxygen and hydrogen groups, and metallic nanoparticles. Then, different chemical modification methods depending on the required products are carried out. This aim may be achieved using wet chemical reactions usually with aqueous solutions of nitric acid, sulfuric acid, hydrochloric acid, oxidation in ozone, oxygen plasma and electrochemical reaction. Functionalized multi walled carbon nanotubes are contained oxygen groups and it is found to be more active than nonfunctionalized multi walled carbon nanotubes. Surface group attached to multi walled carbon nanotubes and surface defects are the active sites for other fictionalizations and catalytic reactions. Obtaining the material with reduced number of defects usually proceeds under annealing.

1.10 FERRI/FERROCYANIDE

Ferrocyanide is the name of the anion $[Fe(CN)_6]^{4-}$ and in aqueous solutions, the coordination number and its complex are fairly unreactive. It is typically available as the salt potassium ferrocyanide which has the formula $K_4Fe(CN)_6$.

The anion of ferrocyanide $[Fe(CN)_6]^{4-}$ is a diamagnetic species featuring the low-spin iron(II) centre in an octahedral ligand environment. Although several salts of cyanides are extremely toxic, ferro and ferricyanides are less toxic due to the high tendency to not release the free cyanide. The most important reaction of ferrocyanide is its oxidation to ferricyanide as follows:

$$[Fe(CN)_6]^{4-} \leftrightharpoons [Fe(CN)_6]^{3-} + e^- \text{ Ferricyanide is the anion } [Fe(CN)_6]^{3-}.$$

Potassium ferricyanide is the chemical compound with the formula $K_3[Fe(CN)_6]$. This bright red salt contains the octahedrally coordinated $[Fe(CN)_6]^{3-}$ ion. Potassium ferricyanide is often used in physiology experiments as a means of increasing a solution's redox potential ($E^{\circ\prime} \sim 436\,mV$ at pH 7). As such, it can oxidize reduced cytochrome c ($E^{\circ\prime} \sim 247\,mV$ at pH 7) in intact isolated mitochondria. Sodium dithionite is usually used as a reducing chemical in such experiments ($E^{\circ\prime} \sim -420\,mV$ at pH 7).

Potassium ferricyanide is used in many amperometric biosensors as an electron transfer agent replacing an enzyme's natural electron transfer agent such as oxygen as with the enzyme glucose oxidase. It is used as this ingredient in many commercially available blood glucose meters for use by diabetics (Bagheri *et al.*, 2015c).

1.11 POLYMERS

1.11.1 Polycaprolactone (PCL)

Biodegradable polymers are increasingly being used as biomaterials. The presentation of such polymeric materials in biological environment is largely due to their surface features.

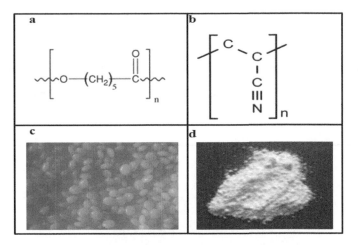

Figure 1.2 Chemical structure of a) Polycaprolactone, b) Polyacrylonitrile, c) Polycaprolactone powder, d) Polyacrylonitrile powder.

Polycaprolactone (PCL) is a biodegradable polymer with low melting point of around 60°C and a glass transition temperature of about −60°C. The molecular formula of PCL is $(C_6H_{10}O_2)_n$ with a density of 1.145 g/cm³. Polycaprolactone can be easily dissolved in water and oil (Figure 1.2 (a and c)). PCL is a semicrystalline linear aliphatic polyester, which is known for its biocompatibility and biodegradability, that make it a useful material in drug delivery systems, and bone graft substitutes. Furthermore, PCL is one of the most hydrophobic commercially available biodegradable polymers. It has good mechanical properties, and it is widely compatible with various types of polymers. It is of major importance to investigate the chain organization on the microscopic scale, especially because it could be a crucial parameter in the determination of potential reactivity and compatibility with other systems.

1.11.2 Polyacrylonitrile (PAN)

Polyacrylonitrile (PAN) is a semicrystalline and synthetic organic polymer with the linear molecular formula $(C_3H_3N)_n$. Although it is thermoplastic, it does not melt under normal conditions. It melts above 300°C and if the heating rates are 50°/min or above. Almost all polyacrylonitrile are copolymers made from the mixtures of monomers with acrylonitrile as the main component. It is a versatile polymer used to produce a large range of products including ultrafiltration membranes, fibers for textiles hollow fibers for reverse osmosis, carbon fibers and oxidized flame retardant fibers. PAN is soluble in many polar organic solvents, such as dimethylformamide (DMF), dimethyl acetemide and dimethyl sulfoxide (Figure 1.2 (b and d)).

1.11.3 Chitosan

Chitosan is a linear co-polymer of Glucosamine and N-acetyl-Glucosamine (Figure 1.3) and it is N-deacetylated derivative of chitin which is naturally occurring bipolymer

Figure 1.3 a) Chemical structure and b) powder of chitosan.

found in the exoskeleton of crustaceans in fungal cell walls and in other biological material.

In this work, chitosan was selected as an inert immobilization matrix in sensor development because it displays an excellent film-forming ability, good adhesion to hydrophilic surfaces, fast ionic transport, high water permeability and ease to chemical modifications due to the presence of reactive amino and hydroxyl groups. Moreover, chitosan is an ideal choice because its solutions contain stable colloidal suspensions of carbon nanotubes which allows for efficient film formation. Chitosan is soluble in acidic solutions in which it behaves as a cationic polyelectrolyte. At pH > 6, the amino groups of chitosan flocculates due to the deprotonation (pKa = 6.5).

1.11.4 Gelatin

Gelatin is one type of denaturation products of collagen and it consists of one single chain of amino acids. Gelatin contains many glycine (almost 1 in 3 residues, arranged every third residue), proline and 4-hydroxyproline residues. A typical structure is -Ala-Gly-Pro-Arg-Gly-Glu-4Hyp-Gly-Pro- (Figure 1.4). It is soluble in warm water and becomes a gel at a concentration higher than 1 wt% (Belton *et al.*, 2004; Dickerson *et al.*, 2004). A gelatin hydrogel is a three-dimensional hydrophilic polymer network which can provide a desirable water-rich buffering environment because of its attractive properties of film forming ability, biocompatibility, non-toxicity, high mechanical strength and cheapness (De Wael *et al.*, 2010). Gelatin is primarily used as a gelling agent forming transparent elastic thermoreversible gels on cooling below about 35°C, which dissolve at low temperature.

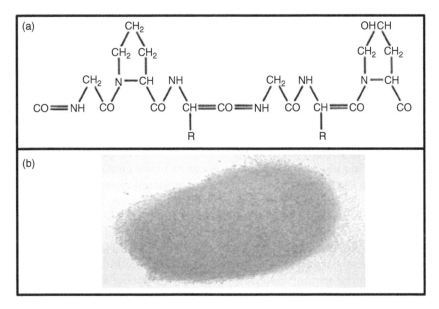

Figure 1.4 a) Chemical structure and b) powder of gelatin.

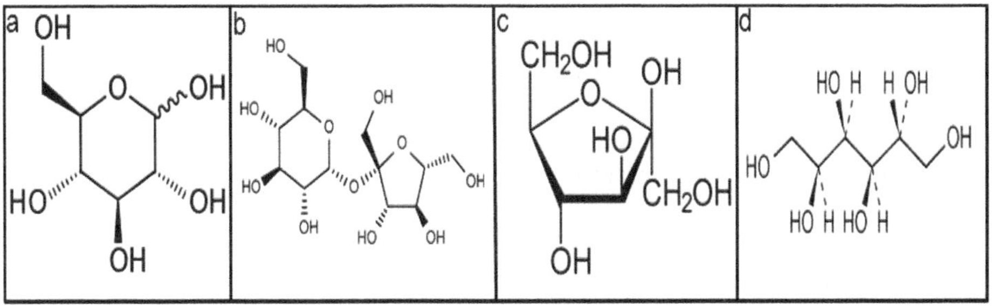

Figure 1.5 Carbohydrates: a) glucose, b) sucrose, c) fructose, d) sorbitol.

1.12 ANALYTES

An analyte is a substance or chemical constituent that is measured. The term is usually applied to a component of blood or another body fluid. In electrochemical sensors discussed in this book, analytes are carbohydrates and amino acids.

1.12.1 Carbohydrates

Carbohydrates are organic compounds that consist of carbon, hydrogen and oxygen. Typically, the hydrogen: oxygen atom ratio of carbohydrate is 2:1 (in water) and with the empirical formula $C_m(H_2O)_n$ (Figure 1.5).

1.12.1.1 Glucose

Glucose is a simple hydrocarbon which exists in plants (Figure 1.5 (a)). It has several different molecular structures but all of these structures can be divided into two families of mirror-images. Glucose is a common medical analyte measured in blood samples. Eating or fasting prior to taking a blood sample has an effect on the result. Higher than usual glucose levels may be a sign of prediabetes or diabetes mellitus.

Glucose monitoring deserves a great deal of attention since diabetes management is of utmost concern for healthcare personnel, individuals and society as it is taking a heavy economic toll. About 285 and 439 million people will be affected by diabetes in 2010 and 2030, respectively (Chekin et al., 2013b). The lifestyle and lifespan of the diabetics are mainly dependent on the accurate blood glucose monitoring, which helps in adjusting the dosage of insulin to be administered so that the glycemic values stays within the normal physiological range. About 85% of the biosensor market belongs to glucose sensors (TermehYousefi et al., 2015c). Therefore, nearly all the sensing concepts are initially tested for the development of glucose sensors as they have a huge commercial impact. Most of the CNTs based electrochemical sensors have also been demonstrated for glucose monitoring applications. Two types of CNTs based electrochemical sensing strategies, based on the presence and absence of enzyme as the bio-recognition element, have been used for the detection of glucose (Chekin et al., 2014a).

1.12.1.2 Sucrose

Sucrose is the organic compound commonly known as table sugar and sometimes called saccharose. An odorless, white and crystalline powder with a sweet taste, it is known due to its nutritional role. It has disaccharide unit composed of hydrocarbon such as glucose and fructose with the molecular formulas $C_{12}H_{22}O_{11}$ (Figure 1.5 (b)).

1.12.1.3 Fructose

Fructose or fruit sugar is a simple hydrocarbon existing in many plants. It is one of the three dietary hydrocarbons along with glucose that is absorbed directly into the bloodstream during digestion. Dry and pure fructose is very sweet, white, odorless, crystalline solid and it is the most water-soluble of all the sugars. From plant sources, fructose is found in honey, tree, vine fruits, flowers, berries and most root vegetables. Fructose is a 6-carbon polyhydroxyketone and it is an isomer of glucose that has the same molecular formula ($C_6H_{12}O_6$) but differs structurally (Figure 1.5 (c)).

1.12.1.4 Sorbitol

Sorbitol ($C_6H_{14}O_6$) is known as glucitol and is a sugar of alcohol which the human body metabolizes slowly (Figure 1.5 (d)). It can be obtained by reduction of glucose and changing the aldehyde group to a hydroxyl group. Sorbitol is found in apples, pears, peaches and prunes.

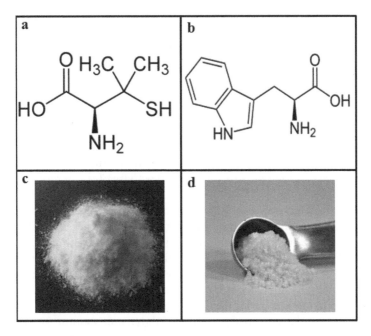

Figure 1.6 Chemical structure of a) D-penicillamine, b) L-Tryptophan, c) D-penicillamine powder, d) L-Tryptophan powder.

1.12.2   D-Penicillamine

Penicillamine ($C_5H_{11}NO_2S$) is a pharmaceutical of the chelator class. It is sold under the trade names of Cuprimine and Depen (Figure 1.6 (a and c)). The pharmaceutical form is D-penicillamine, as L-penicillamine is toxic (it inhibits the action of pyridoxine). It is a metabolite of penicillin, although it has no antibiotic properties.

Bone marrow suppression, dyspepsia, anorexia, vomiting and diarrhea are the most common side effects occurring in ~20–30% of the patients treated with penicillamine.

1.12.3   L-Tryptophan

Tryptophan is ($C_{11}H_{12}N_2O_2$) one of the twenty-two standard amino acids as well as an essential amino acid in the human diet (Figure 1.6 (b and d)). Tryptophan finds medicinal use as an antidepressant and a narcotic as well as in parenteral feeding. The psychogenic effects of tryptophan are caused by its metabolite and the neurotransmitter serotonin, and also the pineal hormone melatonin. It has known sedative effects.

1.13   IMMOBILIZATION TECHNIQUES

Electrochemical sensors and biosensors rely on the use of biologically and chemically active molecules, so the immobilization of recognition elements onto the electrode surface plays a vital role in the development of sensors.

The active molecules are embedded on the electrode surface in a distinct phase that allows the recognition of analytes while separating from the bulk phase, where the analyte molecules are present. Several techniques have been used to immobilize the active molecules onto various electrode surfaces. The frequent immobilization techniques used in this design of electrochemical sensors and biosensors are covalent bonding, entrapment, adsorption, crosslinking and adsorption crosslinking.

In covalent bonding, the recognition element is attached to the active component of electrode surface by means of diverse chemical reacylation such as peptide bond formation or linkage through active functional group such as isocyanato, amino, epoxy, thiol or carboxy of the molecules to be immobilized. Such covalently modified surfaces have the advantages of resistance to changes in temperature, pH and ionic strength. Their disadvantages include the loss of chemical activity of the immobilized molecules due to restricted dynamics.

The entrapment technique is utilized primarily in connection with the development of biosensors or immobilized biocatalysts for heterogeneous catalysis. Mostly, a biomolecule or a recognition element is physically trapped within a polymer matrix or formulated with the components of a membrane prior to laying onto the electrode surface. Typical films of membranes or polymers which are utilized as polyacrylamide, polyvinyl chloride, polyvinyl alcohol, polypyrole and polythiophene. The entrapment technique is a good method and it uses mild conditions and in some of these polymers it is biocompatible and also provides freedom for dynamic motion. The disadvantages include the weak bonding to the surface, leakage of active molecules due to the changes in microstructure inherent to the polymers or swelling properties.

Physical or chemical adsorption is a facile and common method in designing the electrochemical sensors. In this method, the active molecules are attached to the electrode surface by means of hydrogen bonding, ionic bonding, hydrophobic interactions or van der Waals interactions. The major disadvantage of this method is the leakage of the adsorbed reagents as such interactions are susceptible to change in pH, temperature or ionic strength.

The cross linking technique is the most common for the attachment of active molecules onto the electrode surface by means of chemical agents such as hexamethylene, glutaraldehyde, bismaleimidohexane and disuccinyl suberate. Cross linking combines the features of covalent binding, entrapment or adsorption. The disadvantages of this method are the large amounts of active molecules and difficulty in controlling the degree of cross linking.

Most of these immobilization techniques have their own advantages and disadvantages with respect to specific application that are being regularly used in the sensor design.

1.14   ELECTROCHEMICAL METHODS

1.14.1   Cyclic Voltammetry (CV)

The cyclic voltammetric studies are performed by scanning the potential of working electrode in a cyclic manner by applying a potential in one direction and switching it in reverse. The current flowing can be measured through the working electrode.

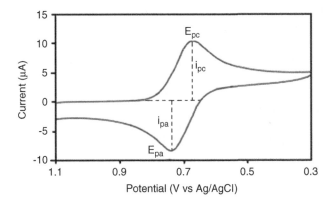

Figure 1.7 Typical cyclic voltammogram.

The plot of the current vs. potential is called cyclic voltammogram (Figure 1.7). The cyclic voltammogram is recorded in stationary solutions that contain an excess of ions of a background electrolyte over the analyte in order to suppress the migration as a mode of the analyte transport to the electrode surface. Under such conditions, the analyte is transported by diffusion to the electrode surface which simplifies the mathematical description of the cyclic voltammograms.

Considering the redox couple B/B_{n-} in Figure 1.7, as the potential of working electrode becomes more negative, species B is reduced to B_{n-}. Ultimately, the concentration of B on the electrode surface is approached zero yield on the cathodic (reduction) current peak (i_{pc}). The current reaches the highest value, at this moment the concentration gradient in the electrode-electrolyte interface is the steepest and the current is proportional to the concentration gradient of species B.

After reaching this peak, the current decreases owing to the concentration gradient decreases due to the spreading of the region of depleted concentration of B further away from the electrode surface (diffusion is too slow to replenish the concentration of B near the electrode surface). After the direction of potential scanning is reversed this scenario is repeated but with B_{n-} being oxidized back to B, which yields the anodic (oxidation) current peak (i_{pa}).

A fully reversible electrode system that is, one of the species (B) having diffusion as the rate determining step and it is characterized by a separation of anodic and cathodic peak potentials at room temperature according to the equation:

$$E_{pa} - E_{pc} = \frac{59}{n} V \tag{1.1}$$

where n is the number of electrons transferred in the reaction. In addition, the mid peak potential between the cathodic and anodic peak potentials approximates so called standard electrode potential ($E°$) of the reversible B/B_{n-} redox system which relates to the standard thermodynamic conditions according to the Nernst equation (25°C):

$$E = E° + \left\{\frac{59}{n}\right\} \log\left(\frac{a_B}{a_{B^{n-}}}\right) \tag{1.2}$$

where a_B and a_{Bn-} are the activities of oxidized and reduced form of species B, respectively. Hence, $E°$ is defined as the electrode potential under standard conditions when all reactants and products are at unit activity. When the activities are replaced with concentrations in Equation 1.2, the standard electrode potential $E°$ becomes the formal (conditional) electrode potential $E_f^°$ which is characteristic for a specific set of experimental condition.

If the electrode reaction is not reversible, that is the diffusion is not rate determining step which indicates the reaction requires an extra driving force (over potential) to proceed at a given rate. The origin of over potential can include the kinetic polarization and concentration polarization. In practical terms, the overpotential shifts the current peaks in the cyclic voltammetry further away from the $E°$, thus the peak separation becomes larger than that calculated value from Equation 1.1.

The generic description of the peak current can be given by:

$$I_p = \frac{dQ}{dt} = \frac{dN}{dt}nF \qquad (1.3)$$

where Q represents the charge in coulombs flowing at time $t(s)$, F is the Faraday constant, N is the number of moles of analyte undergoing the redox reaction at the electrode surface and n represent the number of electrons used in this reaction. From Equation 1.3, one can determine the number of moles of N.

$$N = \frac{Q}{nF} \qquad (1.4)$$

If the reactant is adsorbed on the electrode surface, one can calculate its surface concentration Γ (mol cm^{-2}) by dividing N by the surface area of the electrode.

1.14.2 Electrochemical Impedance Spectroscopy (EIS)

The Electrochemical Impedance Spectroscopy (EIS) is a more general concept of resistance. In direct current (DC) circuits, only resistors oppose the flow of electrons. In alternating current (AC) circuits, the capacitors (impedance) influence the flow of electrons in addition to resistors. Electrochemical impedance spectroscopy is usually measured by applying an AC potential with small amplitude (5 to 10 mV) to an electrochemical cell and measuring the current flowing through the working electrode. The advantage of EIS is that the electrochemical cell can be modeled by using a purely electronic model. An electrode–electrolyte interface undergoing an electrochemical reaction is treated as an electronic circuit consisting of a combination of resistors and capacitors. Frequently, a Randles circuit presented in Figure 1.8 is used. It consists of a solution resistance (R_s) in series with the parallel combination of the double-layer capacitance (C_{dl}) and an impedance of the Faradaic reaction of interest which consists of charge transfer resistance (R_{ct}) and the so called Warburg diffusion element W.

To obtain the Randles circuit parameters, the experimental data are fitted to the model circuit using the non-linear least-squares procedures which are available in the modern EIS software.

Typically, the result of such fitting is presented in the form of the Nyquist plot (Figure 1.8). Such plot displays the kinetic control in the region of high frequencies

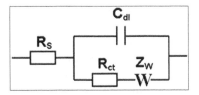

Figure 1.8 Typical fitting Nyquist plot.

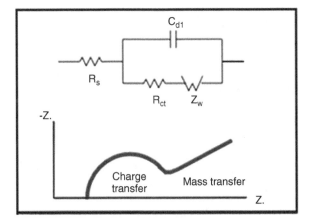

Figure 1.9 Typical fitting according to Nyquist plot.

and a mass transfer (Warburg diffusion) control at low frequencies. The charge transfer resistance of our modified electrodes R_{ct} is determined by measuring the diameter of a semicircle (Figure 1.9), which is proportional to the R_{ct}.

1.14.3 Tafel equation and electron transfer coefficient

The Tafel equation is electrochemical kinetics equation relating to rate of the electrochemical reaction to the overpotential. It was first deduced experimentally and was later shown to have a theoretical justification. The equation is named after Swiss chemist Julius Tafel (1862–1918).

The Tafel equation can be written as equation (1.5) (Scott, 1991):

$$i = nFk \exp\left(\pm \alpha nF \frac{\Delta V}{RT}\right) \tag{1.5}$$

where the plus sign under the exponent refers to an anodic reaction and the minus sign to a cathodic reaction, n is the number of electrons involved in the electrode reaction, k is the rate constant of the electrode reaction, R is the universal gas constant, F is the Faraday constant, T is the temperature, ΔV is the overpotential changes and α is the electron transfer coefficient.

Electron transfer coefficient (α) is a parameter utilized in the description of the kinetics of electrochemical reactions (Marcus, 1964). The electron transfer coefficient signifies the fraction of the interfacial potential at an electrode-electrolyte interface that helps in lowering the free energy barrier for the electrochemical reaction. The electroactive ion present in the interfacial region experiences the interfacial potential and electrostatic work is done on the ion by a part of the interfacial electric field. It is charge transfer coefficient that signifies the part that it is utilized in activating the ion to the top of the free energy barrier.

1.14.4 Butler–Volmer equation

The Butler–Volmer equation is one of the most fundamental relationships in electrochemical kinetics. It describes how the electrical current on the electrode surface depends on the electrode potential and it considers that both a cathodic and anodic reaction occur at the same electrode (Noren & Hoffman, 2005):

$$i = i_o \cdot \left\{ \exp\left[\frac{\alpha_a nF\eta}{RT} \right] - \exp\left[-\frac{\alpha_c nF\eta}{RT} \right] \right\} \tag{1.6}$$

where I is an electrode current (A), i is current density (defined as $i = I/A$), i_o is exchange current density (A/dm^2), T is the absolute temperature (K), n is the number of electrons involved in the electrode reaction, F is the Faraday constant, R is the universal gas constant, α_c is a cathodic electron transfer coefficient, α_a is anodic electron transfer coefficient and η is activation overpotential ($\eta = (E - E_{eq})$).

1.14.5 Mass transfer control and apparent diffusion coefficient

The previous form of the Butler–Volmer equation is valid when the electrode reaction is controlled by electrical charge transfer at the electrode and not by the mass transfer to or from the electrode surface from or to the bulk electrolyte. The usage of the Butler–Volmer equation in electrochemistry is wide, and it is often considered to be central in the phenomenological electrode kinetics.

In the region of the limiting current, when the electrode process is mass-transfer controlled, the value of the current density is:

$$i = \frac{nFD}{\delta} C^* \tag{1.7}$$

where D is the diffusion coefficient, δ is the diffusion layer thickness and C^* is the concentration of the electroactive species in the bulk of electrolyte.

The apparent diffusion coefficient (also referred to as the effective diffusion coefficient) of a diffusant in atomic diffusion of solid polycrystalline materials like metal alloys is often represented as a weighted average of the grain boundary diffusion coefficient and the lattice diffusion coefficient. The ratio of the grain boundary diffusion activation energy over the lattice diffusion activation energy is usually 0.4–0.6, the grain boundary diffusion component increases, as the temperature is lowered.

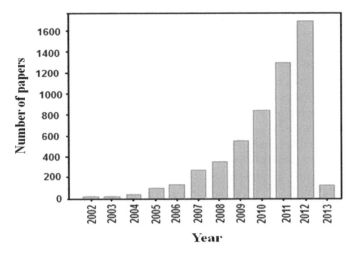

Figure 1.10 Number of publications per year on application of nanomaterials in electrochemical sensors. Data extracted on 15th February 2013 through the Institute of Scientific Information (ISI) database.

1.15 AIM AND OBJECTIVES

According to the previous information and published research papers in the web of science, nanomaterial applications in electrochemical sensors started in early 2000 and has been increasing dramatically since 2006. Figure 1.10 shows the number of publications about this field per year.

Information (ISI) database using the following keyword that appeared in the topic:

application of nanomaterials in electrochemical sensors.

The objectives of this book are:

i)   To synthesize zinc oxide and titanium dioxide nanoparticles by sol-gel method with a suitable, cheap and green reducing polymerizing agent to control the properties of metal oxide nanoparticles.

ii) To fabricate nanocomposite films by using synthesized zinc oxide and titanium dioxide nanoparticles and Multi-Walled Carbon Nanotubes (MWCNTs) for designing electrochemical sensors.

iii) To electrocatalytic oxidation and determine the presence of carbohydrates, *D*-penicillamine and *L*-Tryptophan by using the designed electrochemical sensors.

1.16 BOOK STRUCTURE

This book presents fabrication of nanostructured materials such as metal oxide nanoparticles, carbon nanotubes and nanocomposites and their applications as electrochemical sensors.

Chapter 1 describes the general introduction about nanotechnology, nanoparticles and their application in electrochemical sensors, as well as the aim and objectives of this book.

Chapter 2 presents a literature review on electrochemical sensors, methods of synthesis for zinc oxide and titanium dioxide nanoparticles and fabrication of nanocomposite films.

Chapter 3 discusses materials and techniques that is used to synthesize and characterize nanoparticles as well as a general procedure for fabrication of reported electrochemical sensors.

Chapter 4 describes a novel glucose electrochemical sensor based on electrocatalysis of cobalt(II) immobilized onto the nanocomposite films consist of gel-assisted synthesized zinc oxide nanoparticles, multi walled carbon nanotube and polycaprolactone (ZnO/MWCNT/PCL/GCE). Zinc oxide nanoparticles were characterized by X-ray diffraction technique (XRD), UltraViolet–Visible Spectroscopy (UV-Vis), Fourier Transform Infrared Spectroscopy (FTIR), Scanning Electron Microscopy (SEM) and Transmission Electron Microscopy (TEM).

Chapter 5 describes an electrochemical sensor for electrocatalytic oxidation of L-Tryptophan detection based on synthesized titanium dioxide nanoparticles and multi walled carbon nanotube on glassy carbon electrode (TiO$_2$-MWCNT/GCE). Titanium dioxide nanoparticles were characterized by X-ray diffraction technique (XRD), UltraViolet–Visible Spectroscopy (UV-Vis), Fourier transform infrared spectroscopy (FTIR), Scanning Electron Microscopy (SEM) and Transmission electron microscopy (TEM).

In chapter 6 focus on application of a novel electrochemical sensor for electrocatalytic oxidation of carbohydrates (glucose, sucrose, fructose and sorbitol) detection based on electrocatalysis of nickel (II) into porous polyacrylonitrile- multi walled carbon nanotubes composite modified glassy carbon electrode (Ni/PAN-MWCNT/GCE) was presented. The Electrochemical Impedance Spectroscopy (EIS) and the Scanning Electron Microscopy (SEM) result supported successful immobilization of PAN-MWCNT nanocomposite films.

Chapter 7 reports on application of electrochemical sensor for electrocatalytic oxidation and simultaneous determination of D-penicillamine and L-Tryptophan based on nanocomposite films Containing Chitosan (CS) and Multi Walled Carbon Nanotube (MWCNT) coated on a Glassy Carbon Electrode (GCE). The chitosan films are permeable to anionic $[Fe(CN)_6]^{3-/4-}$ (FC) redox couple. The characterization of modified electrode (CS-MWCNT/GCE) was carried out using scanning electron microscopy (SEM) and UltraViolet–Visible Spectroscopy (UV-Vis).

Chapter 8 presents the book summary and conclusions.

Chapter 2

Literature review of nanocomposites in electrochemical sensors

2.1 SENSORS

Devices with an active sensing material and signal transducer are called sensors. The function of the active sensing material and signal transducer in sensors is to transmit the signal without any magnification from a selective compound or from a change in a reaction (Wilson & Gifford, 2005). Sensors produce any one of the signals as electrical, thermal or optical output signals which could be converted into digital signals for additional processing. One of the methods of categorizing sensors is done based on these output signals (Sakaguchi et al., 2007). Among these methods of classification, the electrochemical sensors have more advantages over the other sensors because in electrochemical sensors, electrodes can sense the materials which are present surrounded by the host without doing any damage to the host system. Moreover, sensors can be generally classified into two categories such as chemical sensors and biosensors (Chen & Chzo, 2006).

The biosensors can be described in terms of sensing aspects, where these sensors can sense biochemical compounds such as biological proteins, tissues and even nucleotides (Vasantha & Chen, 2006). In biosensors, the active sensing material on the electrode surface should act as a catalyst and catalyze the reaction of biochemical compounds to gain the output signals (Simoyi et al., 2003). The mixture of these two different methods of categorizations has given rise to a new type of sensors which are called electrochemical biosensors. In electrochemical biosensors, electrochemical methods are applied to the construction and working of a biosensor (Balasubramanian & Burghard, 2006; Wang et al., 2003; Zhang et al., 2005).

The selection and development of an active material is a challenging task (Yogeswaran et al., 2007a). The active sensing materials can be as a catalyst for sensing a particular analyte or a set of analytes (Yogeswaran & Chen, 2007b). The recent development in nanotechnology has provided the way for a huge number of new materials and devices of attractive properties which have useful functions for various electrochemical sensor and biosensor applications (Hubalek et al., 2007; Yogeswaran & Chen, 2007a; Yogeswaran et al., 2007b). Nanostructure could control the fundamental properties of materials even without changing their chemical composition. In this way the attractive world of low dimensional systems, together with fabrication of functional nanostructured arrays could play a key role in the new trends of nanotechnology (Kim et al., 2006; Liang & Zhuobin, 2003; Lyons & Keeley, 2006).

In addition, the nanostructures can be utilized for both optical excitation and efficient transport of electrons, and these two factors make them vital to the function and integration of nanoscale devices (Hernandez-Velez, 2006; Shie *et al.*, 2008; Yogeswaran & Chen, 2008). Nanosystems are the smallest dimension structures that can be utilized for efficient transport of electrons and are thus vital to the function and integration of these nanoscale devices. Owing to their high surface-to-volume ratio and tunable electron transport features because of quantum confinement effect, their electrical features are influenced by even minor perturbations.

Recent research has illustrated extensive attraction towards nanostructures, mainly on nanoparticles. In general, nanostructures could be synthesized by two methods, one is by "bottom-up" approach where, the self-assembly of small sized structures form into larger structures. The other method is by "top-down" approach where the reduction of large systems down into smaller sizes produces multifunctional nanoscale structures (Li *et al.*, 2005; Vaseashta & Dimova-Malinovska, 2005; Zhang *et al.*, 2005). In the literature, the electrochemical sensor device design and sensing materials selection of appropriate sensing material for appropriate analyte and also their applications are abundant.

2.2   CHEMICAL SENSORS

Electrochemical sensors represent the fastest growing category of chemical sensors. A chemical sensor is a device that provides continuous information about its environment (Janata, 2002). Basically, chemical sensors have a transducer, which transforms the response into a detectable signal on modern instrumentation and a chemically selective layer and which isolates the response of the analyte from its immediate environment. Chemical sensors can be categorized along as per the property to be determined as: optical, thermal, electrical or mass sensors. They are designed to detect and respond to an analyte in solid, gaseous or liquid state. In comparison with optical, thermal and mass sensors, electrochemical sensors are particularly attractive because of their outstanding experimental simplicity, detectability and low cost (Janata & Bezegh, 1988). There are three major types of electrochemical sensors, such as potentiometric, amperometric and conductometric (Widrig *et al.*, 1990).

For potentiometric sensors, a local equilibrium is launched at the sensor interface, where either the membrane or electrode potential is measured, and information about the composition of a sample is obtained from the potential difference between two electrodes. Amperometric sensors employ the use of a potential applied between a working and a reference electrode and to reason the oxidation or reduction of an electroactive species; the resultant current is measured. In contrast, conductometric sensors are engaged with the measurement of conductivity in a series of frequencies. (Janata, 2009; Janata & Bezegh, 1988; Janata & Huber, 1985; Wang, 1988).

2.2.1   Potentiometric sensors

Potentiometric sensors have found the most common practical applicability since the early 1930's, owing to their familiarity, simplicity and cost. There are three basic kinds of potentiometric devices, such as Coated Wire Electrodes (CWES), ion-selective electrodes (IES) and Field Effect Transistors (FETS).

The ion selective electrode is an indicator electrode capable of selectively measuring the activity of a particular ionic species. Classically, these kinds of electrodes are mostly membrane-based devices, consisting of permeable ion-conducting materials, which separate the sample from the inside of the electrode (Heineman *et al.*, 1980). One electrode is the working electrode whose potential is found by its environment. The second electrode is a reference electrode whose potential is fixed with a solution containing the ion of interest at a constant activity (Heller & Yarnitzky, 2001). Because the potential of the reference electrode is fixed, the value of the potential difference (cell potential) can be related to the concentration of the dissolved ion (Bowers & Carr, 1976; Weetall, 1974).

Different methods for producing an electrode that is selective to one species are based primarily on the nature and composition of the membrane material. Research in this area has opened up a variety of applications to an almost limitless number of analytes, where the only limitation is the selection of dopant and ionophore matrix of the membrane (Covington, 1979).

2.2.2 Amperometric sensors

Amperometric sensor is an abnormality in itself. In the context of electroanalytical techniques, the amperometric measurements are made by recording the current flow in the cell at a single applied potential. In addition, a voltammetric measurement is made when the potential difference across an electrochemical cell is scanned from one preset value to another and the cell current is recorded as a function of the applied potential (Wang, 2006).

In both cases, the essential operational feature of voltammetric or amperometric devices is the transfer of electrons to or from the analyte. The basic instrumentation requires controlled-potential equipment and the electrochemical cell compound of two electrodes immersed in a suitable electrolyte. Another complex and common arrangement involves the use of a three-electrode cell, one of the electrodes serving as a reference electrode. While the working electrode is the electrode at which the reaction of interest occurs, the reference electrode such as $Ag|AgCl$, $Hg|Hg_2Cl_2$ provides a stable potential compared to the working electrode. An inert conducting material such as platinum or graphite is typically utilized as counter electrode (Bard & Faulkner, 1980).

A supporting electrolyte is needed in controlled-potential experiments to reduce electromigration effects decrease the resistance of the solution and maintain the ionic strength constant (Brett & Brett, 1993).

The presentation of amperometric sensors is robustly influenced by the working electrode material. Therefore, much effort has been devoted to electrode fabrication and maintenance. While classical electrochemical measurements of analytes started in 1922, when Heyrovsky invented the dropping mercury electrode, for which he received a Nobel Prize, solid electrodes constructed of noble metals and diverse forms of carbon have been the sensors of choice in recent years. The remarkable development in this area, and its growing impact on electroanalytical chemistry, is more recent (Smyth & Vos, 1992).

Mercury was very attractive as an electrode material for many years due to its extended cathodic potential range window, renewable surface and a high reproducibility. The limited anodic potential of mercury electrodes and its toxicity are the principal disadvantages of the method (Kalcher *et al.*, 2005).

Solid electrodes such as carbon, platinum, gold, silver, nickel, copper and dimensionally stable anions have been very popular as electrode materials due to their low background current, versatile potential window, low cost, chemical inertness, and suitability for diverse sensing and detection applications. The explosion of Chemically Modified Electrodes (CME) generates a modern approach to electrode systems, where deliberate alteration of the electrode surface is introduced by incorporation of a suitable surface modifier. Although there was some development in the progress of amperometric sensors in the early 1970s, most amperometric sensors improved were appropriate only to a controlled stringent set of laboratory conditions (Stradiotto *et al.*, 2003).

2.2.3 Conductometric sensors

Chemical sensors in this class relied on changes of electric conductivity of a film or a bulk material, whose conductivity is influenced by the presence of the analyte. Conductimetric methods are basically non-selective (Kriz *et al.*, 1996). Only with the start of modified surfaces for selectivity and much improved instrumentation have these become more feasible methods for designing sensors. There are some very practical considerations that make conductometric methods attractive; simplicity and low cost due to there is no reference electrodes are needed (Svetlicic *et al.*, 1998). Improved instrumentation has contributed to fast and facile determination of analytes, according to the measurement of conductivity (Lesho & Sheppard, 1996).

2.3 CHEMICALLY MODIFIED ELECTRODES

Recently, the immobilization of chemical microstrutures onto electrode surfaces has been a main growth area in electrochemistry (Linford, 1990; Wightman *et al.*, 1989). Chemically Modified Electrodes (CMEs) result from a purposeful immobilization of a modifier agent onto the electrode surface though polymer coating, chemical reactions, composite formation or chemisorption. In comparison with conventional electrodes, greater control of electrode characteristics and reactivity is obtained by surface modification, while the immobilization transfers the physicochemical features of the modifier to the electrode surface (Smyth & Vos, 1992).

Chemically modified electrodes contain development of electrocatalytic systems with high chemical activity and selectivity, coating of semiconducting electrodes with anticorrosive properties and photosensitising, electrochromic displays, microelectrochemical devices in the field of molecular electronics and electrochemical sensors (Wang, 2006).

The main advantages of analytical applications include preferential accumulation, or selective membrane permeation, acceleration of electron transfer reactions and interferent exclusion. Such steps can expose higher selectivity, detectability and stability to amperometric devices (Murray, 1980).

Different kinds of inorganic films, such as metal oxide, clay, zeolite, and metal ferrocyanide, can also be used to modify electrode surfaces (Janek *et al.*, 1998; Pournaghi-Azar & Ojani, 2000). These films are of interest since they frequently indicate well-defined structures, chemically and thermally stable and are typically inexpensive and also readily available (Raoof *et al.*, 2004).

These frequent studies, testify to the reliability of amperometric sensing devices as an electroanaytical method ready to be utilized for solving analytical problems. The wide possibilities of application, as well as high selectivity in pre-concentration and detection, that makes these devices appropriate for analysis of complex samples (Dong & Wang, 1989; Murray *et al.*, 1981).

2.4   CARBON NANOTUBES

Carbon nanotubes have been widely used for electrochemical sensing applications owing to their high surface area, high electric conductivity, ability to accumulate analyte, alleviation of surface fouling, capability to make surface functionalization and electrocatalytic activity (Huang *et al.*, 2003; Lin *et al.*, 2005; Narang *et al.*, 2012; Pumera *et al.*, 2007; Singh *et al.*, 2012; Wang *et al.*, 2002; Zhao *et al.*, 2004; Zhou *et al.*, 2012). Since the discovery of carbon nanotubes in 1991, it has attracted a great amount of attention because of their extraordinary physical, chemical, and mechanical properties (Iijima, 1991). Their large surface area offers rich reactive sites to generate Faradic currents, and the reactions on CNT-modified biosensor electrodes, which belong to the surface controlled process, exhibit direct electron transfer properties (Shamsipur *et al.*, 2010; Tasviri *et al.*, 2011; Zhou *et al.*, 2012).

The development of electrochemical sensors has been widely researched as an inexpensive method to sensitively detect a variety of biological analytes (Kargar *et al.*, 2014). Carbon based electrodes have been commonly used because of their low cost, good electron transfer kinetics and biocompatibility. Recently, carbon nanotubes have also been incorporated into electrochemical sensors. While they have many of the same properties as other types of carbon, CNTs offer unique advantages including enhanced electronic properties, a large edge plane/basal plane ratio, and rapid electrode kinetics (Bagheri *et al.*, 2012a). Therefore, CNT-based sensors generally have higher sensitivities, lower limits of detection, and faster electron transfer kinetics than traditional carbon electrodes (Donya Ramimoghadam, 2015). Many variables need to be tested and then optimized to create a CNT-based sensor. Electrode performance can depend on the synthesis method of the nanotube, CNT surface modification, the method of electrode attachment, and the addition of electron mediators. This review highlights different biomolecules and compares electrode design techniques for selective analyte detection (TermehYousefi *et al.*, 2015a).

The physical and catalytic properties make CNTs ideal for use in sensors. Most notably, CNTs display high electrical conductivity, chemical stability, and mechanical strength. The two main types of CNTs are single-walled CNTs (SWCNTs) and multi-walled carbon nanotubes (MWCNTs). SWCNTs are sp^2 hybridized carbon in a hexagonal honeycomb structure that is rolled into hollow tube morphology. MWCNTs are multiple concentric tubes encircling one another (Mandizadeh *et al.*, 2014).

2.5 ELECTRODES

Electrochemical sensing usually requires a reference electrode, a counter (auxiliary) electrode and a working (sensing or redox) electrode. The reference electrode, usually made from Ag|AgCl is kept at a distance from the reaction site with the aim of maintaining a known and stable potential. The working electrode functions as the transduction element in the biochemical reaction while the counter electrode establishes a connection to the electrolytic solution so that a current can be applied to the working electrode. These electrodes should be both chemically stable and conductive. Thus the platinum, gold, carbon (such as graphite or glass carbon) and silicon compounds are commonly utilized, depending on the analyte.

Amperometric sensors are one type of electrochemical sensor, which they incessantly measure current resulting from the reduction or oxidation of an electroactive species in a biochemical reaction. Usually, the current is measured at a constant potential and this is referred to as amperometry. If a current is measured during controlled variations of the potential this is referred to as voltammetry. Moreover, the peak value of the current measured over a linear potential range is directly proportional to the bulk concentration of the analyte.

2.6 ELECTROCHEMICAL METHODS

2.6.1 Cyclic Voltammetry (CV)

Voltammetry belongs to a group of electro-analytical techniques, through which information about an analyte is gained by varying the potential and then measuring the resulting current. Therefore, voltammetry is an amperometric technique, as there are many ways to vary a potential. There are many forms of voltammetry, such as polarography (DC Voltage) (Heyrovský, 1956), linear sweep, differential pulse voltammetry (DPV), differential staircase, reverse pulse, normal pulse and many more (Katz & Willner, 2003). Cyclic voltammetry is one of the most broadly utilized forms and it is useful to obtain the information about the redox potential and electrochemical reaction rates (such as the chemical rate constant) of analyte solutions. In this case, the voltage is swept between two values at a fixed rate, though, when the voltage reaches V_2 the scan is reversed and the voltage is swept back to V_1. The scan rate, $(V_2 - V_1)/(t_2 - t_1)$, is an important factor, because the duration of a scan must provide adequate time to allow for a meaningful chemical reaction to occur (Wang, 2006).

Thus, varying the scan rate yields correspondingly varied results. The voltage is measured between the reference electrode and the working electrode, whereas the current is measured between the working electrode and the counter electrode. The obtained measurements are plotted as current versus voltage, which is known as voltammogram. While the voltage is raised toward the electrochemical reduction potential of the analyte, the current will also rise. With increasing voltage toward V_2 past this reduction potential, the current decreases, forming a peak as the analyte concentration near the electrode surface diminishes, since the oxidation potential has

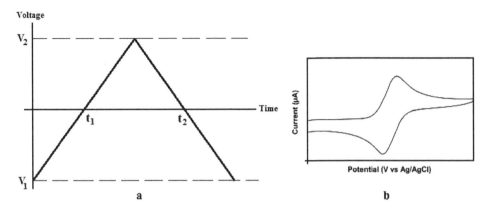

Figure 2.1 a) a single generic linear voltage sweep, b) a typical cyclic voltammogram.

been exceeded. As the voltage is reversed to complete the scan toward V_1, the reaction will begin to reoxidize the product of the initial reaction. This produces an increase in current of opposite polarity as compared to the forward scan, but again a decrease, having formed a second peak as the voltage scan continues toward V_1 (Figure 2.1(a)). The reverse scans also provides information about the reversibility of a reaction at a given scan rate (Gosser, 1993).

The graph of the voltammogram (Figure 2.1(b)) for a given compound relies not only on the scan rate and the electrode surface, which is differentiated after each adsorption step, but can also rely on the catalyst concentration (Liu *et al.*, 2006; Patolsky *et al.*, 1999).

The progress in cyclic voltammetry method is mainly determined by application of a variety of improved electrode materials such as glassy carbon, carbon fibres, carbon nanotubes and conducting polymers (Buratti, 2008).

2.6.2 Chronoamperometry

Another amperometric method is known as chronoamperometry, where the square-wave potential is applied to the working electrode and a steady state current is measured as a function of time (Gosser, 1993).

Changes in the current arise from the expansion or reduction of the diffusion layer at the electrode. The concept of a diffusion layer was introduced by Nernst and states that there is a stationary thin layer of solution in contact with the electrode surface. The local analyte concentration drops to zero at the electrode surface and diffusion controls the transfer of analyte from the bulk solution of higher concentration to the electrode. This results in a concentration gradient away from electrode surface. In the bulk solution the concentration of analyte is maintained at a value of c_0 by convective transfer.

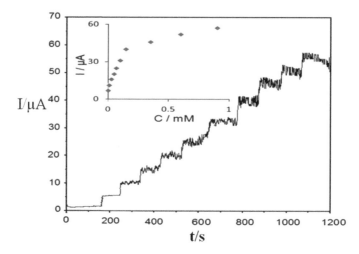

Figure 2.2 Chronoamperometric curve of TiO_2-MWCNT/GCE with successive addition of Tryptophan to a stirred 0.1 M PBS (pH 7.00). The inset is the calibration curve (Chapter 5).

Thus, this technique of chronoamperometry is closely related to the Cottrell equation (Nahir & Buck, 1992):

$$I = nFAc_0 \sqrt{\frac{D}{\pi t}} \tag{2.1}$$

It defines the current-time dependence for linear diffusion control at a planar electrode. In the Cottrell equation, the current I is dependent on Faraday's constant (F), n is the number of transferred electrons per molecule, A is the electrode area; c_0 is the analyte concentration, D is the diffusion coefficient and t is the time. From the Cottrell equation the current depends on the rate at which the analyte diffuses to the electrode.

Figure 2.2 indicates complementary chronoamperometric data on the same system analyzed by chronoamperometry which is often used in addition to other techniques such as cyclic voltammetry, for time-dependent system characterization.

2.6.3 Electrochemical Impedance Spectroscopy (EIS)

The first publication of electrochemical impedance spectroscopy dates to 1975 (Macdonald & Johnson, 2005). Through the application of a small sinusoidally varying potential U, one measures the resulting current response I (Willner & Katz, 2006). By varying the excitation frequency (f) of the applied potential over a range of frequencies, which can calculate the complex impedance, sum of the real and imaginary impedance components, of the system as a function of frequency (angular frequency w). Therefore, the EIS method combines with analysis of both real and imaginary components of impedance, namely the electrical resistance and reactance (Tlili *et al.*, 2006).

EIS has the ability to study any intrinsic material property or specific processes that could influence the conductivity/resistivity or capacitivity of an electrochemical system. Thus, EIS is a useful tool in the improvement and analysis of materials for biosensor transduction, such as the study of polymer degradation (Chang & Park, 2010; Fernández-Sánchez *et al.*, 2005).

For electrochemical sensing, impedance techniques are useful to observe changes in electrical properties arising from biorecognition events at the surfaces of modified electrodes. For example, changes in the conductance of the electrode can be measured as a result of protein immobilization and antibody-antigen reactions on the electrode surface (Bakker, 2004; Janata, 2002). Additionally, to measure the changes in capacitance with EIS sometimes referred to as Faradaic impedance spectroscopy. Binding events of complementary antibody-antigen components alter the electrical properties in the gap between two electrodes, where changes in gap conductivity correspond to changes in the real impedance component $Z_r(w)$ and changes in the gap capacitance correspond to changes in the imaginary impedance component $Z_i(w)$ (Katz & Willner, 2003).

$$Z(jw) = \frac{U(jw)}{I(jw)} = Z_r(w) + jZ_i(w); \quad w = 2\pi f \qquad (2.2)$$

The well-established dynamic methods in electrochemistry such as electrochemical impedance spectroscopy (EIS) and cyclic voltammetry (CV) have obtained more interest in direct sensor applications.

2.7 METAL OXIDE NANOSTRUCTURES

Metal oxides play a very important role in many areas of chemistry, physics, and materials science. The metal elements can form a large diversity of oxide compounds by employing various synthesis techniques (Chekin *et al.*, 2014d). They exhibit metallic, semiconductor, or insulator character due to the electronic structure difference. The variety of attributes of oxides enable the wide applications in the fabrication of microelectronic circuits, sensors, piezoelectric devices, fuel cells, coatings against corrosion, and as catalysts (Bagheri *et al.*, 2015a). For example, almost all catalysts involve an oxide as active phase, promoter (or support) which allows the active components to disperse on. In the chemical and petrochemical industries, products worth billions of dollars are generated every year through processes that use oxide and metal/oxide catalysts (Gholami *et al.*).

Furthermore, the most active areas of the semiconductor industry involve the use of oxides. Thus, most of the chips used in computers contain an oxide component. Till now, there are still many potential applications of these materials under continuous investigation and new synthesis methods being developed. To exploit new applications metal oxide materials is one of the mean purposes of inorganic chemist (Termeh Yousefi, 2015).

From both fundamental and industrial standpoints, the development of systematic methods for the synthesis of metal oxide nanostructures is a challenge, as the first requirement in any study related to oxide nanostructures is the synthesis and characterisation of the material. Methods frequently used for the synthesis of bulk oxides may not work when aiming at the preparation of oxide nanostructures or nanomaterials (Amir *et al.*, 2015). For example, a reduction in particle size by mechanically grinding a reaction mixture can only achieve a limiting level of grain diameter, at best about 0.1 µm. However, chemical methods can be used to effectively reduce particle size into the nanometre range.

2.7.1   Zinc oxide nanoparticles

Zinc oxide (ZnO) is a semiconductor compound with a wide bandgap (3.4 eV) and a stable hexagonal (Wurtzite) structure with lattice spacing $a = b = 0.325$ nm and $c = 0.521$ nm. Table 2.1 shows crystallographic parameters of ZnO (ICDD card number: 01-079-0206).

Due to its exclusive properties and versatile applications in ultraviolet light emitters, transparent electronics, chemical sensors, piezoelectric devices, and spin electronics, it has attracted intensive research attempts (Espitia *et al.*, 2012). Based on outstanding physical properties and the motivation of device miniaturization, great effort has been focused on the synthesis, characterization and device applications of ZnO nanomaterials (Ellmer *et al.*, 2008).

2.7.1.1   Synthesis of zinc oxide nanostructures

There are lots of synthesis methods for preparing zinc oxide nanoparticles such as precipitation (Meulenkamp, 1998), thermal decomposition (Lin & Li, 2009), hydrothermal synthesis (Kuo *et al.*, 2005; Liu & Zeng, 2003; Zhang *et al.*, 2004; Zhou, *et al.*, 2007), sol-gel (Niederberger, 2007; Rani *et al.*, 2008; Tokumoto *et al.*, 2003; Vafaee & Ghamsari, 2007) and others. Some characteristics of these methods such as their precursor as well as size and shape of the obtained nanoparticle are presented in Table 2.2. In the present research work we report the synthesis of zinc oxide and titanium dioxide nanoparticles by sol-gel technique.

Table 2.1 Crystallographic parameters of ZnO (ICDD card number: 01-079-0206).

Structure	Lattice parameter (a, b, c (nm))	2θ (deg)	hkl	d_{hkl} (nm)	Intensity (%)
Hexagonal	$a = b = 0.325$	31.768	(100)	0.281	57.8
	$c = 0.521$	34.422	(002)	0.260	41.9
		36.253	(101)	0.247	100.0
		47.539	(102)	0.191	21.4
		56.594	(110)	0.162	30.3
		62.858	(103)	0.147	26.5
		66.374	(200)	0.140	4.0
		67.947	(112)	0.137	21.3
		69.085	(201)	0.135	10.2
		72.568	(004)	0.130	1.6
		76.959	(202)	0.123	3.2
		81.387	(104)	0.118	1.6
		89.613	(203)	0.109	6.4

Table 2.2 Synthesis methods of ZnO nanoparticles.

Method	Precursor	Solvent	Size (nm)	Shape	Reference
Coprecipitation technique	Zinc acetate	Double distilled water	80 (length), 30–60 (diameter)	Nanorod	(Bhadra, et al., 2011)
Microwave decomposition	Zinc acetate dehydrate	1-butyl-3-methylimidazolium bis (trifluoromethylsulfonyl) imide	37–47	Sphere	(Jalal et al., 2010)
Hydrothermal process	Zinc acetate	Polyvinylpyrrolidone (PVP)	5 μm (length), 50–200 (diameter)	Nanorod	(Elen et al., 2009)
Wet chemical method	Zinc nitrate	Sodium hydrate (NaOH) as precursor	20–30	Acicular	(Espitia et al., 2012)
Sol-gel method	Zinc nitrate	Distilled water	30–60	Circular and hexagonal	(Zak et al., 2011)

Generally, preparation of nanoparticles is a complicated process, and different variables may affect the properties of the final product. Some significant variables have distinct effects on the properties of the final product, while others may have only minor effects. In the case of synthesis of zinc oxide nanoparticle, variables such as nanoparticle size should be controlled in order to obtain a uniform size distribution (Meulenkamp, 1998).

2.7.2 Titanium dioxide nanoparticles

The nanosized titanium dioxide (TiO_2) particles have attracted lots of interest to materials scientists and physicists due to their special properties and have attained a great importance in several technological applications such as photocatalysis, sensors, solar cells and memory devices (Fujishima *et al.*, 2000).

Titanium dioxide exists in three polymorphic phases, such as rutile (tetragonal density $= 4.25$ g/cm^3), anatase (tetragonal, 3.894 g/cm^3) and brookite (orthorhombic, 4.12 g/cm^3) (Table 2.2 and 2.3 show the crystallographic parameters of anatase and rutile). Both anatase and rutile have tetragonal crystal structures but belong to different space groups. Anatase has the space group I4$_1$/amd (Zhang, 2006) with four TiO_2 formula units in one unit cell and rutile has the space group P4$_2$/mnm (Lin, 2006) with two TiO_2 formula units in one unit cell (Zhang, 2000). The low-density solid phases are less stable and undergo transition to rutile phase in the solid state. The transformation between anatase and rutile is accelerated by heat treatment and occurs at temperatures between 450 and 1200°C (Tsevis, 1998). This transformation relies on several parameters such as initial particle size, dopant concentration, initial phase, reaction atmosphere and annealing temperature (Diebold, 2003).

Table 2.3 Crystallographic parameters of anatase (ICDD card number: 01-073-1764).

Structure	Lattice parameter (a, b, c (nm))	2θ (deg)	hkl	d_{hkl} (nm)	Intensity (%)
Anatase	a = b = 0.377	27.126	(101)	0.350	100
	c = 0.948	35.581	(103)	0.242	5.7
		38.738	(004)	0.237	18.0
		40.686	(112)	0.232	6.6
		43.530	(200)	0.188	23.0
		53.608	(105)	0.169	14.0
		55.943	(211)	0.166	13.5
		61.760	(213)	0.148	2.2
		63.255	(204)	0.147	9.8
		64.640	(116)	0.136	4.1
		68.080	(220)	0.133	4.5
		68.705	(107)	0.127	0.4
		71.439	(215)	0.126	6.6
		73.439	(301)	0.124	1.8
		75.334	(008)	0.118	0.3
		78.569	(303)	0.116	0.4
		81.180	(224)	0.116	3.1
		83.105	(312)	0.115	1.2

Table 2.4 Crystallographic parameters of rutile (ICDD card number: 01-076-0324).

Structure	Lattice parameter (a, b, c (nm))	2θ (deg)	hkl	d_{hkl} (nm)	Intensity (%)
Rutile	a = b = 0.464	27.126	(110)	0.328	100
	c = 0.300	35.581	(101)	0.252	42.9
		38.738	(200)	0.232	6.0
		40.686	(111)	0.221	16.0
		43.530	(210)	0.207	5.8
		53.608	(211)	0.170	40.6
		55.943	(220)	0.164	10.2
		61.760	(002)	0.150	5.5
		63.255	(310)	0.146	4.5
		64.640	(221)	0.144	0.3
		68.080	(301)	0.137	10.6
		68.705	(112)	0.136	5.8
		71.439	(311)	0.131	0.5
		73.439	(320)	0.128	0.2
		75.334	(202)	0.126	1.1
		78.569	(212)	0.121	0.6
		81.180	(321)	0.118	1.5
		83.105	(400)	0.161	1.0
		86.270	(410)	0.126	0.4
		88.098	(222)	0.110	2.6
		89.425	(330)	0.109	1.2

2.7.2.1 Crystal structure of titanium dioxide

Among the three above mentioned crystal structures of titanium dioxide, the anatase phase has the highest photocatalytic activity and it is usually utilized for photocatalysis

(Bickley *et al.*, 1991). This higher photocatalytic activity is related to its lattice structure. Each titanium atom is coordinated to six oxygen atoms in anatase tetragonal unit cell (Ohsaka *et al.*, 2005). Yong *et al.* (Liang *et al.*, 2001) have reported a significant degree of buckling associated with O-Ti-O bonds in anatase compared to rutile titanium dioxide. Crystal symmetry is reduced owing to this buckling, and in turn it results in a larger unit cell dimensions in the (001) direction. The kinetics of phase transformations in titanium dioxide is extensively reviewed by Zhang *et al.* (Zhang *et al.*, 2008). Figure 2.3 shows the unit cell structures of the anatase and rutile crystals. The titanium and oxygen atoms are more tightly packed in the rutile crystal. Both structures can be described in terms of chains of titanium dioxide octahedra. Each Ti^{+4} ion is surrounded by an octahedron of six O^{-2} ions. The octahedron in rutile is not regular, indicating a slight orthorhombic distortion.

The octahedron in anatase is considerably distorted so that its symmetry is lower than orthorhombic. The Ti-Ti distance in anatase is longer while the Ti-O distances are shorter than in rutile. In the rutile structures each octahedron is in contact with ten neighbouring octahedrons (two sharing edge oxygen pairs and eight sharing corner oxygen atoms) whereas in the anatase structure each octahedron is in contact with eight neighbours (four sharing an edge and four sharing a corner). These differences in lattice structures cause different mass densities and electronic band structures between the two forms of titanium dioxide as indicated in Figure 2.3. Anatase can be conceived as an arrangement of parallel octahedral, while in case of rutile some octahedral are rotated by 90°. There is a symmetry change from I4$_1$/amd to P4$_2$/mnm space group in terms of reconstructive polymorphism during conversion from anatase to rutile. Consequently, the ionic mobility that occurs during phase transition results in increased densification and coarsening of the titanium dioxide nanoparticles (Diebold, 2003).

Titanium dioxide is stable in aqueous media and is tolerant of both alkaline and acidic solutions. It is recyclable, reusable, inexpensive and relatively simple to produce. It can also be synthesized in nanostructured forms easier than many other catalysts.

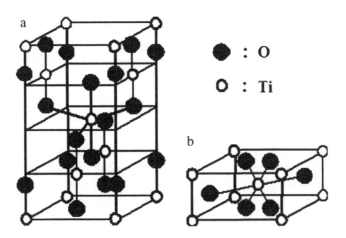

Figure 2.3 Unit cell structures of the a) anatase and b) rutile crystals.

Moreover, its bandgap is appropriate to initiate a variety of organic reactions (Carp *et al.*, 2004).

2.7.2.2 *Synthesis of titanium dioxide nanostructures*

Titanium dioxide nanoparticles can be synthesized by various methods ranging from simple chemical to mechanical and also to vacuum methods, including many variants of physical and chemical vapour deposition methods (Trentler *et al.*, 1999). In the present research work the synthesis of titanium dioxide nanoparticles by sol-gel method is reported.

Titanium dioxide nanoparticles can be synthesized using various methods such as sulphate process (Li *et al.*, 2005), chloride process (Pratsinis & Spicer, 1998; West *et al.*, 2007), impregnation (Uguina *et al.*, 1994), precipitation (Ranjit & Viswanathan, 1997; Tang *et al.*, 2005), hydrothermal method (Corradi *et al.*, 2005; Puddu *et al.*, 2007; Suzuki & Yoshikawa, 2004; Yin *et al.*, 2001), direct oxidation of $TiCl_4$ (West *et al.*, 2007), metal organic chemical vapour deposition (Ding *et al.*, 2001). Sol-gel method (Lakshmi *et al.*, 1997; Subrahmanyam, *et al.*, 2012; Wang & Ying, 1999; Zhang *et al.*, 2001) is one of the most convenient ways to synthesize a variety of metal oxides owing to low cost, ease of fabrication and low processing temperatures.

It is broadly used to prepare titanium dioxide for films, particles or monoliths. Generally, the sol gel method involves the transition of a system from a liquid sol into a solid gel phase. The homogeneity of gels relies on the solubility of reagents in the solvent, the sequence of addition of reactants, temperature and the pH. The precursors normally used for the synthesis and doping of nanoparticles are organic alkoxides, acetylacetonates or acetates as well as inorganic salts such as chlorides (Almquist & Biswas, 2002). The sol-gel method is based on inorganic polymerization reaction. It includes four steps, hydrolysis, polycondensation, drying and decomposition. Hydrolysis of the precursors (metal or non-metal alkoxides) takes place with water or alcohols (Brinker & Scherer, 1990).

$$Ti(OR)_4 + 4H_2O \rightarrow Ti(OH)_4 + 4ROH \qquad \text{(hydrolysis)}$$

where R is the alkyl group. In addition to water and alcohol, an acid or a base also helps in the hydrolysis of the precursor. After condensation of the solution of a gel, the solvent is removed (Brinker & Scherer, 1990).

$$Ti(OH)_4 \rightarrow TiO_2 + 2H_2O \qquad \text{(condensation)}$$

Calcination at higher temperature is needed to decompose the organic precursor. The size of the sol particles depends on the solution composition, pH and temperature. By controlling these factors, one can tune the size of the particles. The major advantages of sol-gel processing are ambient temperature of sol preparation and gel processing, low temperature of sintering, ease of making multi-component materials, product homogeneity, and good control over powder particle size, shape and size distribution (Klein, 1994).

2.8 NANOCOMPOSITES

A nanocomposite is a multiphase solid material where one of the phases has one, two or three dimensions of less than 100 nanometers (nm), or structures which have nano-scale repeated distances between the different phases that make up the material (Ajayan *et al.*, 2006). Generally, this definition can include porous media, gels, colloids and copolymers, but it is more commonly taken to mean the solid combination of a bulk matrix and nano-dimensional phase(s) differing in properties owing to dissimilarities in structure and chemistry.

The electrical, thermal, mechanical, optical, catalytic and electrochemical properties of the nanocomposite will differ noticeably from that of the component materials. Size limits for these effects have been proposed, <5 nm for catalytic activity, <20 nm for making a hard magnetic material soft, <50 nm for refractive index changes, and <100 nm for achieving superparamagnetism, mechanical strengthening or restricting matrix dislocation movement (Advani, 2007). In mechanical terms, nanocomposites differ from conventional composite materials because of remarkably high surface to volume ratio of the reinforcing phase and/or its exceptionally high aspect ratio (Ajayan *et al.*, 2006).

This large amount of surface area means that a relatively small amount of nanoscale reinforcement can have an observable effect on the macroscale properties of the composite. For example, adding carbon nanotubes improves the electrical and thermal conductivity (Moniruzzaman & Winey, 2006). Nanoparticulates could result in enhanced optical properties, dielectric properties, heat resistance or mechanical properties such as stiffness, strength and resistance to wear and damage (Zhan *et al.*, 2003). Generally, the nano reinforcement is dispersed into the matrix during processing. The mass fraction (percentage by weight) of the nanoparticles introduced can remain very low (in the order of 0.5% to 5%) because of the low filler percolation threshold, especially for the most commonly used non-spherical, high aspect ratio fillers such as nanometer-thin platelets, for example clays, or nanometer-diameter cylinders, such as carbon nanotubes (Vilgis *et al.*, 2009).

One of the main nanocomposites is carbon nanotube matrix composites which is emerging as a new materials that are being developed to take advantage of the high electrical conductivity and tensile strength of carbon nanotube materials (Singh *et al.*, 2012). Critical to the realization of carbon nanotube matrix composites possessing optimal properties in these areas are the development of synthetic methods that are economically producible, providing a homogeneous dispersion of nanotubes in the matrix, and lead to strong interfacial adhesion between metallic matrix and the carbon nanotubes (Ajayan *et al.*, 2006).

2.9 SOLVENT DISPERSION AND CASTING IMMOBILIZATION

The most broadly used techniques for fabricating carbon nanotubes based electrochemical sensors involve dispersing of carbon nanotubes in a solvent with sonication, after their purification and activation pre-treatments, followed by dropping the resultant suspensions on the electrode surfaces and allowing it to dry (such as the casting methods). Among the reported solvents, N,N-dimethylformamide (DMF) is the most extensively used polar solvent and more than half of the papers deal with carbon nanotubes based electrochemical sensors use DMF as the dispersing solvent (Jianxiu Wang et al., 2002). Other solvents utilized to prepare carbon nanotubes suspensions include water (Wang et al., 2002), acetone (Gojny et al., 2003), ethanol (Li et al., 2004) and even toluene (Lou et al., 2004). However, compared with these solvents, DMF has some overwhelming advantages for dispersing carbon nanotubes. For example, DMF gave higher solubility, stability, and exfoliation efficiency. Other methods of dispersing the carbon nanotubes suffer from a number of disadvantages, such as low exfoliation efficiency, low solubility and low stability because of the rather weak interactions between these solvents and carbon nanotubes. To improve the solubility and stability of CNTs in their suspensions, various additives are added into solvents to assist the dispersion of CNTs, such as surfactants and polymers.

2.10 CHITOSAN

Jiang et al. reported preparation method for a stable dispersion of carbon nanotubes in acidic aqueous solutions of chitosan with sonication and used to simultaneous determination of D-ascorbic acid and uric acid (Jiang et al., 2004). The special interaction between chitosan and carbon nanotubes was characterized by Zhang et al. (Zhang et al., 2004). They demonstrated using thermogravimetric analysis (TGA) that chitosan could be adsorbed onto carbon nanotubes and form a special chitosan-carbon nanotubes system, which can subsequently precipitated from the solutions by the addition of concentrated salts or the adjustment of solution acidity.

The selective interaction between chitosan and carbon nanotubes also provided a possible approach for separating carbon nanotubes from carbonaceous impurities. Based on the functionalization of the reactive groups on chitosan, they developed a carbon nanotubes based glucose electrochemical biosensor that might be applied to a large group of dehydrogenase enzymes for designing a variety of bioelectrochemical devices such as sensors, biosensors and biofuel cells (Star, et al., 2002).

Jiang et al. (Jiang et al., 2004) applied the chitosan-carbon nanotubes system for the direct determination of nitrite (Jiang et al., 2005), the simultaneous detection of D-ascorbic acid and uric acid (Lu et al., 2005), and the selective determination of D-ascorbic acid in the presence of uric acid (Jiang et al., 2005). Another amazing work on the simultaneous electrodeposition of chitosan-carbon nanotubes on gold electrodes was reported by Lue et al. (Luo et al., 2005).

2.11 DETERMINATION OF CARBOHYDRATES

Electrocatalytic oxidation of carbohydrates is of great interest from medical applications of the blood glucose sensing, to ecological approaches like food industry, and waste-water treatment. Most previous studies about determination of carbohydrates involved the use of glucose oxidase, which catalyzes the oxidation of glucose to gluconolactone (Ye & Baldwin, 1994). Owing to the intrinsic nature of enzymes, such enzyme-based sensors suffer from stability problem (Zhang et al., 2005). Therefore, great efforts are made to achieve direct determination of carbohydrates. Recently, Ye et al. reported the direct electrocatalytic oxidation of glucose in alkaline medium with well-aligned multi walled carbon nanotubes electrodes (Ye et al., 2004).

Furthermore, the amperometric responses of these species with the carbon nanotube modified electrodes were extremely stable, with no loss in sensitivity after continuous several hours of operation. Besides unmodified carbon nanotubes, the composites made up of other nanoparticles and carbon nanotubes were also used for electrooxidation of carbohydrates. Male *et al.* reported a copper decorated carbon nanotube based electrochemical sensor for the determination of carbohydrates (Male *et al.*, 2004).

2.11.1 Determination of glucose

Recently, nanostructured materials are used to improve the stability and sensitivity of glucose sensors (Meng *et al.*, 2009). Detection of glucose with electrochemical sensors by using glucose oxidase exhibits high selectivity and sensitivity, but it lacks stability owing to inherent brittleness of the enzyme (German; Yildiz *et al.*, 2005). The main drawback of these enzymatic based glucose sensors is their lack of stability originating from the nature of the enzyme, which is difficult to overcome (Liang & Zhuobin, 2003; Liu *et al.*, 2009).

To address this problem, many attempts have been made to develop glucose sensors without using enzymes. Several nanostructured materials have been reported, and their novel characteristics certainly provide new opportunities to develop innovative nonenzymatic glucose sensors (Dong *et al.*, 2012; Male *et al.*, 2004). For instance, glucose detection has been reported utilizing mesoporous platinum where its mesoporous surface retains sufficient sensitivity. Nanotubular platinum array electrodes possess high selectivity and sensitivity, due to their high surface-roughness structure (Baic *et al.*, 2008). Copper oxide nanowire array electrodes exhibit high sensitivity, selectivity, and immunity to chloride poisoning (Luo, Zhu, & Wang). The immobilization of carbon nanotubes and nanocomposites including platinum, platinum-lead alloy, nickel, gold, and copper, on the electrode surface has also been an important strategy in the construction of nonenzymatic glucose sensors (Bai *et al.*, 2008; Liang & Zhuobin, 2003; Zhang *et al.*, 2011).

2.12 DETERMINATION OF AMINO ACIDS

Amino acids are usually electroactive and commonly contain phenol moieties, so, they can be detected by electrochemical methods (Meister, 1957). Xu *et al.* prepared a multi walled carbon nanotubes film modified electrode and coupled this electrode to an ion chromatography system for the simultaneous determination of oxidizable amino acids, including cysteine, tryptophan and tyrosine (Xu *et al.*, 2003). Wu *et al.* investigated the electrochemical behavior of tryptophan at multi walled carbon nanotubes modified glassy carbon electrode (Wu *et al.*, 2004). The presence of multi walled carbon nanotubes boost the oxidation peak current of tryptophan, and induced shift of the peak potential negatively. Under the chosen conditions, the differential pulse voltammetry peak current was linear to the concentration of tryptophan in the range of $2.5 \times 10^{-7}-1.0 \times 10^{-4}$ M, and the detection limit was 2.7×10^{-8} M. The voltammetric response of tryptophan at a single wall carbon nanotubes modified glassy carbon electrode was also studied by Huang *et al.* (Huang *et al.*, 2004).

Chapter 3

Experimental procedures and materials for nanocomposites in electrochemical sensors

3.1 INTRODUCTION

One of the key successes to any investigation depends on its experimental methodology and reliability of data obtained. In this chapter the experimental setup used to produce nanoparticles and nanocomposites via a sol-gel method, various analytical and spectroscopic characterization techniques and different technological applications are presented in brief. The results must be reproducible and these depend on the experimental procedures and materials employed during the course of study. Well-planned and executed experiments commonly give reproducible and reliable data.

The main experimental steps used in the present investigation are connected to synthesis, characterization and applications of nanosized metal oxides by sol-gel method and preparation of nanocomposites by drop-casting technique and electrochemical measurements. All these experimental steps are different from each other and therefore in each case standard procedures were adopted for collecting the results. Many variable factors and parameters like concentration of analyte, current density, pH, calcination temperature, sample purity, surface uniformity of glassy carbon electrode, the nature of surface modification by agitation etc are adjusted to get accurate results. Many times duplicate experiments were conducted to confirm the results of the particular experiment. All the results recorded in this book are reproducible within the given experimental limitations.

In the present work maximum care is taken in conducting the experiments and collecting data and wherever necessary, some experiments were repeated. The following materials and methods are adopted in the present investigation.

3.1.1 Materials

In this book, AR grade chemicals with high purity are used as listed in Table 3.1. Multi walled carbon nanotube (outer diameter: 13–16 nm, length: 1–10 μm, purity >95 and arc discharge method) were purchased from Baytubes Co. Ltd. (Germany). In this work, de-ionized water (18.2 MΩ-cm at 25°C) was used for preparing all solutions and also was used to clean all materials and glasswares.

Table 3.1 List of chemical compounds used.

Chemical compounds	Molecular formula	Brand	Purity (%)
Zinc nitrate	$Zn(NO_3)_2$	Sigma Aldrich	≥ 99.5
Cobalt(II) nitrate	$Co(NO_3)_2$	Sigma Aldrich	>99.0
Polycaprolactone	$(C_6H_{10}O_2)_n$	Sigma Aldrich	≥ 97.5
Polyacrylonitrile	$(C_3H_3N)_n$	Sigma Aldrich	>97.5
L-Tryptophan	$C_{11}H_{12}N_2O_2$	Sigma Aldrich	>97.5
D-penicillamine	$C_5H_{11}NO_2S$	Sigma Aldrich	≥ 97.0
Chitosan	$(C_6H_{11}O_4N)n$	Sigma Aldrich	≥ 97.5
Potassium ferrocyanide	$K_4[Fe(CN)_6]$	Sigma Aldrich	≥ 98.0
Potassium ferricyanide	$K_3[Fe(CN)_6]$	Sigma Aldrich	≥ 98.0
Sodium hydroxide	$NaOH$	Sigma Aldrich	>97.5
Potassium chloride	KCl	Sigma Aldrich	>97.5
Acetic acid	$C_2H_4O_2$	Sigma Aldrich	>99.0
Titanium isopropoxide	$Ti[OCH(CH_3)_2]_4$	Acros Organics	>98.0
Glucose	$C_6H_{12}O_6$	Merck	≥ 97.5
Sucrose	$C_{12}H_{22}O_{11}$	Merck	≥ 97.5
Fructose	$C_6H_{12}O_6$	Merck	≥ 97.5
Sorbitol	$C_6H_{14}O_6$	Merck	≥ 97.5
Nitric acid	HNO_3	Fluka	>89.0
Sulfuric acid	H_2SO_4	Fluka	>88.0
N,N-dimethyl formamide	C_3H_7NO	Fluka	>90.0
Orthophosphoric acid	H_3PO_4	Fluka	≥ 97.5
Monosodium phosphate	NaH_2PO_4	Fluka	≥ 98.0
Disodium phosphate	Na_2HPO_4	Fluka	≥ 97.5
Trisodium phosphate	Na_3PO_4	Fluka	≥ 99.0

3.2 SYNTHESIS OF NANOCRYSTALLINE METAL OXIDES BY SOL-GEL METHOD

Metal oxides play a very important role in many areas of chemistry, physics and materials science, and often exhibit enhanced physical, chemical, thermal, electrical, optical or magnetic properties, which lead to the extensive applications in electrochemistry, biomedical device and other fields. In this context, preparation of metal oxide particles is a very important task for material scientists. In this work, we have synthesized the metal oxide nanoparticles by the sol-gel method.

The solution chemistry method, usually referred to sol-gel method, is a significant process to synthesize the precursors of many nanoscale metal oxides. The method is widely used for it can achieve uniform doping of multi-elements no matter whether at atomic, molecular or nanometer levels at the gelatination phase. Generally, sol-gel processes have associated problems, such as difficulty in removing chlorine, and in accurate and repeated doping.

The sol-gel method was established by Geffcen and Berger from Shott Company established the way of achieving the sol-gel process for oxide layers (Brinker & Scherer, 1990). In recent years, nanosized metal oxide powder and nanocomposites are synthesized by sol-gel method because of its superior homogeneity, purity, low temperature operation process, good mixing for multi-component systems and effective control of particle size, shape, and properties (Hench & Ulrich, 1986). In the sol-gel method, typical precursors are metal alkoxides and metal chlorides which undergo various forms

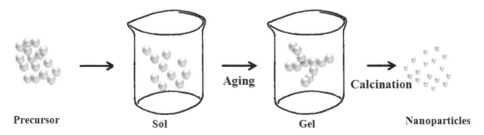

Figure 3.1 Schematic diagram of sol-gel method for the synthesis of metal oxides.

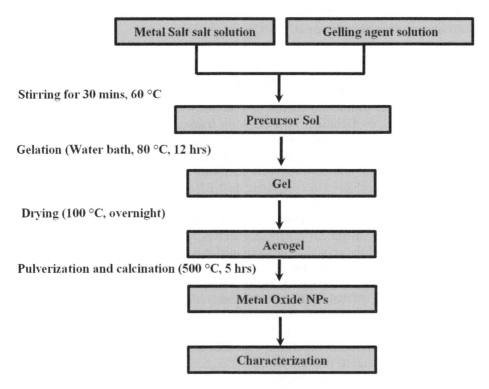

Figure 3.2 Work flow scheme for the synthesis of metal oxide nanoparticles by sol-gel method.

of hydrolysis and polycondensation reactions. The formation of a metal oxide involves connecting the metal centers with either oxo (M–O–M) or hydroxo (M–OH–M) bridges and generating metal-oxo or metal-hydroxo polymers in solution. Thus, the sol evolves towards the formation of a gel-like system containing both liquid and solid phases whose morphologies range from discrete particles to continuous polymer networks (Pierre, 1998). The schematic diagram of the sol-gel process is shown in Figure 3.1.

Figure 3.2 shows the work flow scheme for the synthesis of metal oxide nanoparticles by sol-gel method.

3.3 CHARACTERIZATION OF SYNTHESIZED METAL OXIDE PARTICLES

The successful chemical synthesis of nanocomposite materials can be judged by the yield and the homogeneity of the product, and also ensuring that the materials performance by using analytical and spectroscopic techniques. The characterization of nanocomposite materials is an important part in the determination of shape, size, phase purity, chemical composition, functional group and optical band gap. In the present work, the as-synthesized particles are calcined at different temperature and structurally characterized by different analytical and spectroscopic techniques like powder X-ray diffraction (XRD), thermogravimetric analysis (TGA), scanning electron microscopy (SEM)/energy-dispersive X-ray spectroscopy (EDAX), transmission electron microscopy (TEM), Fourier transform infrared spectroscopy (FTIR) and ultraviolet–visible spectroscopy (UV-Vis).

3.3.1 Powder X-ray Diffraction (XRD)

The powder X-ray diffraction technique was discovered by German Scientist W.C. Roentgen in 1895, the technique is based on the measurement of fluorescence, absorption, and scattering and diffraction of X-rays has been developed and broadly used to investigate the compositions and structures of matters. X-ray Diffraction (XRD) is a versatile and powerful non-destructive technique for characterizing crystalline materials. Identification is achieved by comparing the XRD pattern or diffractogram obtained from an unknown sample with an internationally recognized database containing reference patterns for more than 148,000 phases (Warren, 1990).

X-ray diffraction method is widely utilized for characterization of composite materials. In this method the crystal to be examined is reduced to a very fine powder and placed in a beam of monochromatic X-rays. Each particle of powder is a tiny crystal oriented randomly with respect to the incident beam. The result is that certain sets of lattice planes will be capable of producing reflection (Klug & Alexander, 1974). It provides the information about the amorphous content of the sample, crystalline size, crystallinity, solid solutions, stress and texture. This technique is a well-established tool to confirm the formation of solid state reaction, observe the impurity phases, determine the lattice constants, interplanar distances, octahedral and tetrahedral site radii, etc (Suryanarayana & Norton, 1998).

Modern computer controlled diffractometer systems use automatic routines to measure record and interpret the unique diffractogram produced by individual constituents in even highly complex mixtures. Figure 3.3 shows the D8 Advance X-Ray Diffractometer. It has the ability to perform very rapid phase identifications of powdered specimens in a fully automated mode.

Data collection and manipulation is under control of EVA software which contains a database of the JCPDS powder diffraction files. Diffraction spectra are plotted and can be compared (in whole or in the selected portions) to specified JCPDS cards or automatically matched to the most similar spectra in the database. This instrument operates very rapidly and efficiently. It is capable of very rapid scans for identification of powders that are simple in composition and abundance on the mount. Much slower

Figure 3.3 X-Ray Diffractometer (Bruker D8 Advance).

scans are usually required for the analysis of complex mixed phased, high resolution works, cell refinement or identification of trace impurities.

The average crystallite sizes can be determined by using the Debye-Scherrer's equation:

$$D = \frac{K\lambda}{\beta_{hkl} \cos\theta} \tag{3.1}$$

where, D is the crystal size perpendicular to the reflection plane, K is unit cell geometry dependent constant whose value is typically between 0.85–0.99, λ is the wavelength of X-rays provided by the equipment used (1.5418 Å), β_{hkl} is the broadening of the diffraction line measured at half the line maximum intensity (full width at half maxima) and θ is the angle between the incoming X-rays and the reflection plane ($2\theta/2$).

The Bragg β_{hkl} can be written as

$$\beta_{hkl} = (B_{hkl} - b) \tag{3.2}$$

where, B_{hkl} is the broadening measured on the X-ray pattern, and b is the unavoidable broadening error of the equipment, which is typical of each instrument. However, the X-ray powder diffraction spectra have been used to determine lattice constants and to confirm single-phase nature, phase purity and crystal structure of synthesized metal oxides.

3.3.1.1 Measuring conditions

The XRD analysis was performed using a Bruker X-ray diffraction model D-8 (made in Germany) equipped with EVA diffraction software for data acquisition and analysis. Data were acquired using CuK_α monochromatized radiation using source operated at 40 kV and 40 mA at ambient temperature. The samples were finely grinded and placed in the sample holder, with the powder lightly passed into place using microscopic slide and the surface of samples flattened and smoothened. The sample holder was placed in the diffractometer stage for analysis. A continuous 2θ scan mode from 2°–90° was used for high degree scanning at step time of 4s and step size of 0.01° 2θ. A divergence slit was inserted to ensure that the X-rays focused only on the sample and not the edges of specimen holder. The diffractograms obtained were matched against the Joint Committe on Powder Diffraction Standards (JCPDS) PDF 1 database 2.6 to confirm the identity of samples.

3.3.2 Thermogravimetric Analysis (TGA)

Thermogravimetric Analysis (TGA) is an analytical technique where the properties of materials are studied as they change with temperature (Menczel & Prime, 2009). It is widely used to evaluate heat resistance and to determine the quantity of components, weight loss and thermal effects during the conversion of precursors to final metal oxides in the heat-treatment process. The schematic diagram of TGA is shown in Figure 3.4.

TGA is one of the most versatile and well known instruments for measurement of thermal effect in R&D and QC environments. TGA measures the weight of a substance subjected to a temperature program. Our TGA offers latest balance technology and automatic weight calibration. With the thermobalance, small weight losses can be measured and to identify a weight step, the weight change should be at least twice as large as the peak-to-peak noise, about 2 μg.

A TGA consists of a sample pan that is supported by a precision balance. That pan resides in a furnace and is heated or cooled during the experiment. The mass of the sample is monitored during the experiment. A sample purge gas controls the sample environment. This gas may be inert or a reactive gas that flows over the sample and exits through an exhaust.

These instruments can quantify loss of water, loss of solvent, loss of plasticizer, decarboxylation, pyrolysis, oxidation, decomposition, weight % filler, amount of metallic catalytic residue remaining on carbon nanotubes, and weight % ash. All these quantifiable applications are usually done upon heating, but there are some experiments where information may be obtained upon cooling.

3.3.2.1 Measurement conditions

To investigate the thermal stability of the samples, the measurement was done using Thermogravimetric Analysis (TGA). The TGA analysis was carried out using Mettler

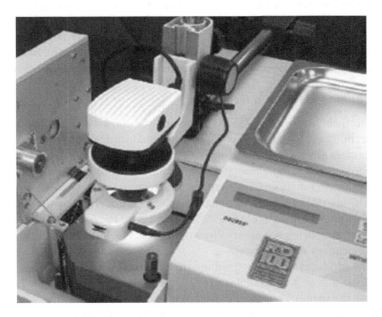

Figure 3.4 Schematic diagram of thermal analysis (Mettler Toledo (SDTA-85 1e)).

Toledo (SDTA-85 1e) thermobalance. About 10 to 50 mg of sample was put into an open alumina crucible of 100 μl for the measurements. The TGA temperature program was run dynamically from ambient to 548 K at a heating rate 10 K min^{-1} under nitrogen gas with flow rate 50 ml min^{-1}. It was cooled to room temperature before being heated again under Argon gas from ambient to 873 K at a heating rate 2 K min^{-1} with Argon gas at the same flow rate.

3.3.3 Electron microscopy

An electron microscope is a microscope that uses a beam of accelerated electrons as a source of illumination. Because the wavelength of an electron can be up to 100,000 times shorter than that of visible light photons, the electron microscope has a higher resolving power than a light microscope and can reveal the structure of smaller objects. A transmission electron microscope can achieve better than 50 pm resolution and magnifications of up to about 10,000,000x whereas most light microscopes are limited by diffraction to about 200 nm resolution and useful magnifications below 2000x. The transmission electron microscope uses electrostatic and electromagnetic lenses to control the electron beam and focus it to form an image. These electron optical lenses are analogous to the glass lenses of an optical light microscope.

Electron microscopes are used to investigate the ultrastructure of a wide range of biological and inorganic specimens including microorganisms, cells, large molecules, biopsy samples, metals, and crystals. Industrially, the electron microscope is often used for quality control and failure analysis. Modern electron microscopes produce electron micrographs using specialized digital cameras and frame grabbers to capture the image.

Electron microscopy is extremely versatile for providing structural information over a wide range of resolution from 10 μm to 0.2 nm. Especially in the range where the specimen is so small (<1 μm) that optical microscopes are not able to image it anymore. Electron microscopes operate in either reflection or transmission mode. The basic principle and setup of each working mode are briefly discussed below.

3.3.3.1 Scanning electron microscopy

Scanning Electron Microscopy (SEM) is one of the most versatile and well known analytical techniques. Compared to the conventional optical microscope, an electron microscope offers advantages including high magnification, large depth of focus, great resolution and ease of sample preparation and observation. Electrons generated from an electron gun enter a surface of a sample and generate many low energy secondary electrons. The intensity of these secondary electrons is governed by the surface topography of the sample (Goldstein *et al.*, 2003). An image of the sample surface is therefore constructed by measuring secondary electron intensity as a function of the position of the scanning primary electron beam.

The SEM produces images by probing the specimen with a focused electron beam that is scanned across a rectangular area of the specimen. When the electron beam interacts with the specimen, it loses energy by a variety of mechanisms. The lost energy is converted into alternative forms such as heat, emission of low-energy secondary electrons and high-energy backscattered electrons, light emission or X-ray emission, all of which provide signals carrying information about the properties of the specimen surface, such as its topography and composition. The image displayed by an SEM maps the varying intensity of any of these signals into the image in a position corresponding to the position of the beam on the specimen when the signal was generated. In the SEM image of an ant shown at right, the image was constructed from signals produced by a secondary electron detector, the normal or conventional imaging mode in most SEMs.

In addition to secondary electrons imaging, backscattered electrons imaging and Energy Dispersive X-ray (EDAX) Analysis are also useful tools widely used for chemical analysis. The intensity of backscattered electrons generated by electron bombardment can be correlated to the atomic number of the element within the sampling volume. Hence, qualitative elemental information can be revealed. The characteristic X-rays emitted from the sample serve as fingerprints and give elemental information of the samples including semi-quantitative analysis, quantitative analysis, line profiling and spatial distribution of elements (elemental/chemical mapping). SEM with X-ray analysis is efficient, inexpensive, and non-destructive to surface analysis. It has wide ranges of applications both in industry and research.

3.3.3.2 Transmission electron microscopy

Being indispensable for nanotechnology, High Resolution Transmission Electron Microscopy (HRTEM) is one of the most powerful tools used for the determination of shape and structure of nanomaterials and will play a critical role in this field in the future.

TEM technique is introduced to the analysis of transmitted or forward-scattered beam. An electron beam interaction with a solid specimen results in number of elastic or inelastic scattering phenomena. Such a beam is passed through a series of lenses,

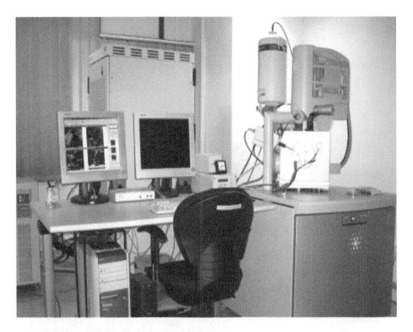

Figure 3.5 Scanning electron microscopy layout (FEI quanta 200F).

among which the objective lens mainly determines the image resolution, to obtain the magnified image (Williams & Carter, 2009). The schematic diagram of transmission electron microscopy (TEM) is as shown in Figure 3.6.

In TEM, the ray of electrons is produced by current heating of a pin-shaped cathode. The electrons are vacuumed up by a high voltage at the anode. The accelerating voltage is between 50 and 150 kW. The higher it is, shorter are the electron waves and higher is the power of resolution. The accelerated beam of electrons passes a drill-hole at the bottom of the anode. Its following way is analogous to that of a ray of light in a light microscope. The beam is first focused by a condenser and then passes through the object, where it is partially deflected. The degree of deflection depends on the electron density of the object. The greater the mass of the atoms, the greater is the degree of deflection. After passing the object the scattered electrons are collected by an objective lens.

An important mode of TEM utilization is electron diffraction. The advantages of electron diffraction over X-ray crystallography are that the specimen need not be a single crystal or even a polycrystalline powder, and also that the Fourier transform reconstruction of the object's magnified structure occurs physically and thus avoids the need for solving the phase problem faced by the X-ray crystallographers after obtaining their X-ray diffraction patterns of a single crystal or polycrystalline powder. The major disadvantage of the transmission electron microscope is the need for extremely thin sections of the specimens, typically about 100 nanometers. Biological specimens are typically required to be chemically fixed, dehydrated and embedded in a polymer resin to stabilize them sufficiently to allow ultrathin sectioning. Sections of biological

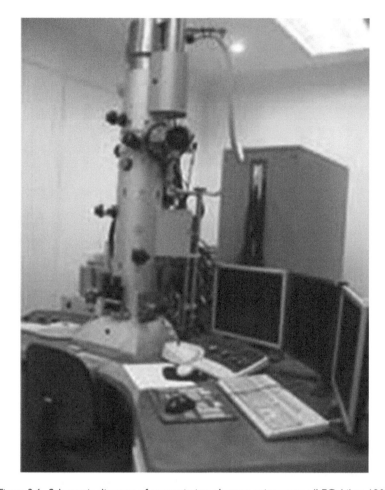

Figure 3.6 Schematic diagram of transmission electron microscopy (LEO-Libra 120).

specimens, organic polymers and similar materials may require special treatment with heavy atom labels in order to achieve the required image contrast.

Thereby an image is formed, that has subsequently been enlarged by an additional lens-system. Thus formed image is made visible on a fluorescent screen or it is documented on photographic material.

In the present study, LEO-Libra 120 (Germany) transmission electron microscope is used to study the morphology of the products.

3.3.3.3 Measurement conditions

The surface morphology of the samples was studied by using a Quanta 200 FEI FESEM instrument, shown in Figure 3.5 and EXAD was used to analyze the chemical composition of the sample surface. The powder samples were adhered to the aluminium sample holder using a small piece of carbon conductive tape before loaded to the sample chamber of the instrument. The air evacuation in chamber was performed before analysis. The EDAX were analyzed using INCA-Suite version 4.02 software. The detailed parameters of the measurement were as follows:

Accelerating voltage:	5 kV
Vacuum mode:	High vacuum
Working distance:	8–10 mm
Spot size:	2.5–3 μm
Type of detector:	Large Field Detector (LFD)
Cone:	X-ray

For EDAX, the acceleration voltage was dispersed on a carbon tape adhered to an aluminium sample holder. Two types of detectors were used, i.e. secondary electron (Evehart-Thornley) and backscattered electron (Robinson).

The TEM specimens were prepared by dropping the samples on a copper grid. A LEO-Libra 120 microscope operated at 200 kV and equipped with a field emission gun. For the size measurement and phase identification images of representative areas of the sample were taken at different magnifications with a CCD camera.

3.3.4   Fourier transform infrared spectroscopy

Fourier Transform Infrared Spectroscopy (FTIR) is an analytical technique used to identify organic (and in some cases inorganic) materials (Griffiths & De Haseth, 2007). This technique measures the absorption of various infrared light wavelengths by the material of interest. These infrared absorption bands identify specific molecular components and structures. The early-stage IR instrument is of the dispersive type, which uses a prism or a grating monochromator. The dispersive instrument is characteristic of a slow scanning (Ferraro & Basile, 1975). A Fourier Transform Infrared (FTIR) spectrometer obtains infrared spectra by first collecting an interferogram of a sample signal with an interferometer, which measures all of infrared frequencies simultaneously. An FTIR spectrometer acquires and digitizes the interferogram, performs the FT function, and outputs the spectrum (Smith, 2009).

Infrared (IR) spectroscopy is a chemical analytical technique, which measures the infrared intensity versus wavelength (wave number) of light. Based upon the wave number, infrared light can be categorized as far infrared ($4\sim400\,cm^{-1}$), mid infrared ($400\sim4,000\,cm^{-1}$) and near infrared ($4,000\sim14,000\,cm^{-1}$). Infrared spectroscopy detects the vibration characteristics of chemical functional groups in a sample (Siesler, Ozaki, Kawata, & Heise, 2008). When an infrared light interacts with the matter, chemical bonds will stretch, contract and bend. As a result, a chemical functional group tends to adsorb infrared radiation in a specific wave number range regardless of the structure of the rest of the molecule. Figure 3.7 shows the FTIR spectroscopic instrument. Therefore, IR spectroscopy can result in a positive identification (qualitative analysis) of every different kind of material.

FTIR spectrometers are mostly used for measurements in the mid and near IR regions. For the mid-IR region, 2–25 μm (5000–$400\,cm^{-1}$), the most common source is a silicon carbide element heated to about 1200 K. The output is similar to a blackbody. Shorter wavelengths of the near-IR, 1–2.5 μm (10000–$4000\,cm^{-1}$), require a higher temperature source, typically a tungsten-halogen lamp. The long wavelength output of

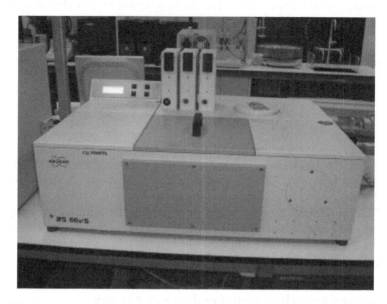

Figure 3.7 Schematic diagram of Fourier transform infrared spectroscopy (Bruker, IFS 66v/s).

these is limited to about $5\,\mu m$ ($2000\,cm^{-1}$) by the absorption of the quartz envelope. For the far-IR, especially at wavelengths beyond $50\,\mu m$ ($200\,cm^{-1}$) a mercury discharge lamp gives higher output than a thermal source.

Mid-IR spectrometers commonly use pyroelectric detectors that respond to changes in temperature as the intensity of IR radiation falling on them varies. The sensitive elements in these detectors are either deuterated triglycine sulfate (DTGS) or lithium tantalate (LiTa). These detectors operate at ambient temperatures and provide adequate sensitivity for most routine applications. To achieve the best sensitivity the time for a scan is typically a few seconds. Cooled photoelectric detectors are employed for situations requiring higher sensitivity or faster response. Liquid nitrogen cooled mercury cadmium telluride (MCT) detectors are the most widely used in the mid-IR. With these detectors an interferogram can be measured in as little as 10 milliseconds. Uncooled indium gallium arsenide photodiodes or DTGS are the usual choices in near-IR systems. Very sensitive liquid-helium-cooled silicon or germanium bolometers are used in the far-IR where both sources and beam splitters are inefficient.

An ideal beam-splitter transmits and reflects 50% of the incident radiation. However, as any material has a limited range of optical transmittance, several beam-splitters may be used interchangeably to cover a wide spectral range. For the mid-IR region the beam splitter is usually made of KBr with a germanium-based coating that makes it semi-reflective. KBr absorbs strongly at wavelengths beyond $25\,\mu m$ ($400\,cm^{-1}$) so CsI is sometimes used to extend the range to about $50\,\mu m$ ($200\,cm^{-1}$). ZnSe is an alternative where moisture vapour can be a problem but is limited to about $20\,\mu m$ ($500\,cm^{-1}$). CaF_2 is the usual material for the near-IR, being both harder and less sensitive to moisture than KBr but cannot be used beyond about $8\,\mu m$ ($1200\,cm^{-1}$). In a simple Michelson interferometer one beam passes twice through the beam splitter but the other passes through only once. To correct for this an additional compensator plate of equal thickness is incorporated. Far-IR beam splitters are mostly based on polymer films and cover a limited wavelength range.

3.3.4.1   Measurement conditions

FTIR spectrum was recorded in the transmittance mode using a Bruker-optic spectrometer IFS 66v/s equipped with vacuum using the technique of KBr pellets and working with a resolution of $4\,cm^{-1}$ in the Middle range. Before analysis, air evacuation was done under vacuum (5 mbar) for 15 minutes. All the spectra recorded a 32 scan data accumulation in a range 400–4000 cm^{-1}.

3.3.5   UV–Visible spectroscopy

Ultraviolet–visible spectroscopy or ultraviolet–visible spectrophotometry (UV-Vis or UV/Vis) refers to absorption spectroscopy or reflectance spectroscopy in the ultraviolet–visible spectral region. This means it uses light in the visible and adjacent (near-UV and near-infrared [NIR]) ranges. The absorption or reflectance in the visible range directly affects the perceived color of the chemicals involved. In this region of the electromagnetic spectrum, molecules undergo electronic transitions. This technique is complementary to fluorescence spectroscopy, in that fluorescence deals with transitions from the excited state to the ground state, while absorption measures transitions from the ground state to the excited state.

When light (UV or Visible) is absorbed by valence electrons, these electrons are promoted from their normal states to higher energy excited states. When light passes through the compound, energy from the light is used to promote an electron from a bonding or non-bonding orbital into one of the empty anti-bonding orbital (Giusti & Wrolstad, 2001). Each wavelength of light has the particular energy associated with it. If that particular amount of energy is just right for making energy jumps, then that wavelength will be absorbed, its energy will have been used in promoting an electron. That means, in order to absorb light in the region from 200–800 nm, the molecule must contain either pi bonds or atoms with non-bonding orbitals. Nonbonding orbital is a lone pair, say oxygen, nitrogen or a halogen. In general as the size of the materials is reduced to nanodimension, blue shift is observed from the UV-Vis spectrum. The schematic diagram of UV–Visible spectroscopy (UV-Vis) is as shown in Figure 3.8. In the present study absorption spectrum of the sample is measured on UV–Visible spectrophotometer (Perkin Elmer, Lambda 35).

3.3.5.1   Measurement conditions

Ultraviolet–visible (UV-vis) spectroscopy is used to obtain the absorbance spectra of a compound in solution. The observed spectra are the absorbance of light energy or electromagnetic radiation, which excites electrons from the ground state to the first singlet excited state of the compound or material. The UV-Vis region of energy for the electromagnetic spectrum covers 1.5–6.2 eV which relates to a wavelength range of 800–200 nm.

Figure 3.8 Schematic diagram of UV–Visible spectroscopy (Perkin Elmer, Lambda 35).

3.4   CHEMICALLY MODIFIED ELECTRODES (CMEs)

3.4.1   Pre-treatment of the electrodes

The fundamental process in electrochemical reactions is the transfer of electrons between the working electrode surface and molecules in the interfacial region (either in solution or immobilized at the electrode surface). The kinetics of this heterogeneous process can be significantly affected by the microstructure and roughness of the electrode surface, the blocking of active sites on the electrode surface by adsorbed materials, and the nature of the functional groups (e.g., oxides) present on the surface. Therefore, there has been considerable effort devoted to finding methods that remove adsorbed species from the electrode and produce an electrode surface that generates reproducible results.

The most common method for surface preparation is mechanical polishing. The protocol used for polishing depends on the application for which the electrode is being used, and the state of the electrode surface. There are a variety of different materials available (e.g., diamond, alumina, silicon carbide), with different particle sizes suspended in solution. The pad used for polishing also depends on the material being used for polishing pads are used with alumina polish, and nylon pads should be used with diamond polish. New working electrodes have first been lapped to produce a flat surface, and have then been extensively polished to a smooth, mirror-like finish at the factory. Therefore, they typically only require repolishing with 0.05 mm or 1 mm diamond polish by the user in between experiments. The electrode should be moved in a figure-of-eight motion when polishing to ensure uniform polishing. Materials that have a rougher surface must first be polished using a larger particle polish, in order to remove the surface defects. After the defects have been removed, the polishing should continue with successively smaller particle size polish.

Once polishing has been completed (this can require from 30 s to several minutes, depending upon the state of the electrode), the electrode surface must be rinsed thoroughly with an appropriate solvent to remove all traces of the polishing material (since its presence can affect the electron transfer kinetics). Alumina polishes should be rinsed with distilled water, and diamond polishes with methanol or ethanol. The rinsing solution should be sprayed directly onto the electrode surface. After the surface has been rinsed, electrodes polished with alumina should also be sonicated in distilled water for a few minutes to ensure complete removal of the alumina particles. If more than one type of polish is used, then the electrode surface should be thoroughly rinsed between the different polishes.

As discussed above, the effect of any surface pretreatment can be determined by its effect on the rate of electron transfer. This can be judged qualitatively by examining the separation of the peak potentials in a cyclic voltammogram of a molecule whose electron transfer kinetics are known to be sensitive to the state of the surface (a more quantitative determination can be made by calculating the value of the standard heterogeneous rate constant k_s from this peak potential separation). For example, k_s for potassium ferricyanide at a glassy carbon surface following a simple polishing protocol is typically in the range 0.01–0.001 cm s^{-1} (this should be compared with the values measured for k_s for a platinum electrode, which are at least one order of magnitude larger). The strong dependence of the electron transfer kinetics of ferricyanide on the state of the electrode surface means that there can be significant variations in the peak potential separation after each polishing, since polishing alters the microstructure, roughness, and functional groups of the electrode surface in addition to removing adsorbed species. The materials used for the polishing can also affect the value of k_s. For example, the electrode surface can be contaminated by the agglomerating agents required to keep the alumina particles suspended in solution and by the components of the polishing pad. The presence of these species can have a deleterious effect on the electron transfer kinetics by blocking the active sites for the electron transfer reaction. However, it should be noted that such pronounced dependence on the state of the electrode surface is only observed for certain systems (the most well characterized examples are the reduction of ferricyanide, the oxidation of ascorbate, and the adsorption of dopamine). For such systems, polishing is often used in combination with another pretreatment (e.g., heat or electrochemical). However, for many other systems, the simple polishing described above is adequate (for example, when using non-aqueous electrolytes, since blocking of active sites by adsorbed species is less common in such electrolytes than it is in aqueous electrolytes).

3.4.2   Preparation of phosphate buffer

Phosphate buffered solutions (PBS) 0.1 M for different pH values were prepared from orthophosphoric acid (H_3PO_4) and its salts (NaH_2PO_4, Na_2HPO_4 and Na_3PO_4). Phosphate buffer solutions containing potassium chloride at a concentration of 0.1 M used as the carrier electrolyte to stabilize the pH range of pH 2 to pH 12 were used in the test solutions. The following solutions were used for different pH ranges:

pH 2 to pH 4: From a solution of phosphoric acid and monosodium phosphate (NaH_2PO_4).

pH 5 to pH 9: From a solution of monosodium phosphate (NaH_2PO_4) and disodium hydrogen phosphate (Na_2HPO_4).

pH 10 to pH 12: From a solution of disodium hydrogen phosphate (Na_2HPO_4) and trisodium phosphate (Na_3PO_4).

3.4.3   Preparation of serum samples and real sample analysis

For determination of glucose in human blood serum, the serum sample was collected from university of Malaya hospital. Around $50\,\mu L$ of serum sample was transferred to the cell continuing 10 mL of 0.1 M NaOH solution. The response of the modified sensor into mentioned solution was investigated by cyclic voltammetric measurement at a scan rate of $50\,mV\,s^{-1}$.

For the purpose of proving its practical applications, the modified electrode was used to determine the content of D-Penicillamine (D-PA) in the tablet using Differential Pulse Voltammetry (DPV). A standard addition method was adopted to estimate the accuracy and the measurement.

3.5   FABRICATION OF CHEMICALLY MODIFIED ELECTRODES

Pre-treatment or cleaning of electrode surface prior to modification is one of the important steps in the fabrication process and in electrochemical studies.

3.5.1   Fabrication of Co/ZnO/MWCNT/PCL/GCE electrode

About 1% of polycaprolactone (PCL) solution was prepared by dissolving the appropriate amount of PCL into Dimethyl Formamide (DMF) with the aid of ultrasonic agitation until complete dissolution. Appropriate amount of ZnO nanoparticles and MWCNTs were dispersed in 1% of PCL solution. The mass ratio of ZnO:MWCNT: PCL was 1:2:20. The mixture was ultrasonicated for about 20 min. Finally, a highly dispersed black colloidal solution was formed. About $5\,\mu L$ of ZnO/MWCNT/PCL casting solution was coated on the GCE surface and dried in air. Then, the obtained ZnO/MWCNT/PCL/GCE was immersed into 1.0 mM cobalt nitrate solution for 5 min at room temperature to obtain Co/ZnO/MWCNT/PCL/GCE, followed by washing the electrode with water to remove the unimmobilized Co(II) ions. Figure 3.9 shows the fabrication of Co/ZnO/MWCNT/PCL/GCE electrode.

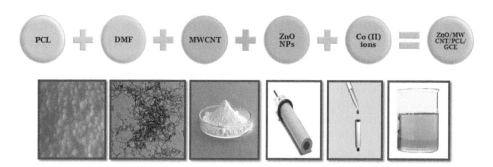

Figure 3.9   Fabrication of Co/ZnO/MWCNT/PCL/GCE electrode.

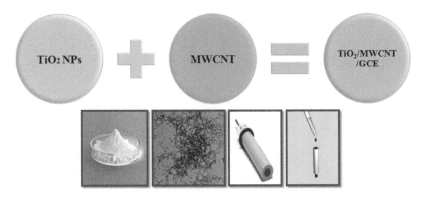

Figure 3.10 Fabrication of TiO$_2$-MWCNT/GCE electrode.

3.5.2 Fabrication of TiO$_2$-MWCNT/GCE electrode

Appropriate amount of gelatin based TiO$_2$ nanoparticles and MWCNTs were dispersed in Dimethyl Formamide (DMF) solution. The mass ratio of TiO$_2$:MWCNT was 1:5. The mixture was ultrasonicated for 20 min. Finally, a highly dispersed black colloidal solution was formed. Prior to use, the glassy carbon electrode (diameter 2 mm) was first polished with alumina (0.05 μm) slurry and it was ultrasonically cleaned with ethanol followed by double distilled water then dried at room temperature. Figure 3.10 shows the fabrication of TiO$_2$-MWCNT/GCE electrode.

3.5.3 Fabrication of Ni/PAN-MWCNT/GCE electrode

About 1.0 mg of MWCNT was dispersed in 5 mL of N,N-dimethyl formamide (DMF) with the aid of ultrasonic agitation to give a black dispersion. On the other hand PAN (1.0 mg) was dissolved in 1.0 mL of DMF. After that, the PAN and MWCNT solutions with a volume ratio of 1:1 were mixed and then the mixture was agitated in an ultrasonic bath for 30 min to form a uniform mixture. The cleaned and dried GCE was coated by casting 5 μL of the black PAN-MWCNT suspension and dried at 50°C in hot air oven to remove the solvent. Then the PAN-MWCNT/GCE was immersed in 1.0 mM nickel nitrate solution for 30 min to obtain Ni/PAN/MWCNT/GCE composite film. For comparison, Ni/GCE, Ni/PAN/GCE and Ni/MWCNT/GCE were prepared using similar procedure as described above. Figure 3.11 shows the fabrication of Ni/PAN-MWCNT/GCE electrode.

3.5.4 Fabrication of FC/CS-MWCNT/GCE electrode

Around 5 mg of chitosan was dissolved in 1 mL 1% acetic acid, and after which pH of the solution was adjusted to pH 5 with concentrated NaOH. With the aid of ultrasonic agitation, 1 mg of MWCNTs was dissolved in 1 mL of 0.5 wt. % CS solution and resulted in a homogeneous black CS-MWCNT. 5 μL of the resulting solution was then casted onto the glassy carbon electrode surface to prepare a CS-MWCNT/GCE

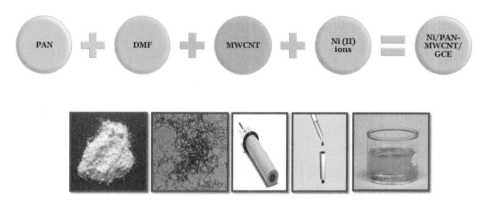

Figure 3.11 Fabrication of Ni/PAN-MWCNT/GCE electrode.

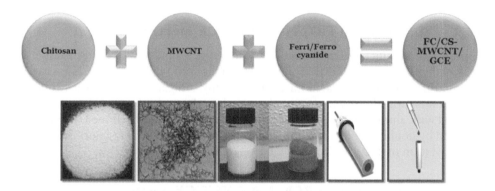

Figure 3.12 Fabrication of FC/CS-MWCNT/GCE electrode.

electrode then dried at room temperature. CS-MWCNT/GCE electrode was placed in an aqueous solution containing 1 mM $[Fe(CN)_6]^{3-/4-}$ and 0.1 M KCl. The electrochemical behavior of the $[Fe(CN)_6]^{3-/4-}$ redox couple was probed by cycling the potential of the CS-MWCNT/GCE electrode between 0.00 V and 0.45 V at a scan rate of 0.05 V/s for 40 complete cycles. CS-MWCNT/GC electrode was later removed from the $[Fe(CN)_6]^{3-/4-}$ solution, rinsed and were cycled in the same potential window in supporting electrolyte (0.1 M KCl) to probe for residual $[Fe(CN)_6]^{3-/4-}$ at the FC/CS-MWCNT/GCE electrode. Then for comparison, FC/GC, FC/CS/GCE and FC/MWCNT/GCE electrodes were prepared with the same procedure as described above. Figure 3.12 shows the fabrication of FC/CS-MWCNT/GCE electrode.

3.6 ELECTROCHEMICAL MEASUREMENTS OF MODIFIED ELECTRODES

3.6.1 Electrochemical setup

The electrochemical measurements were performed using a three-electrode system including Working Electrode (WE), Reference Electrode (RE), and Counter (auxiliary)

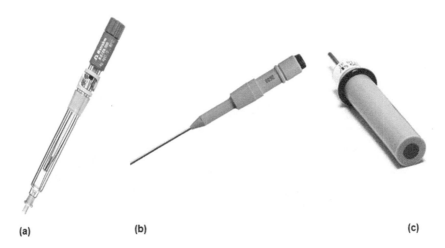

(a)   (b)   (c)

Figure 3.13 Three-electrode system a) reference electrode, b) counter electrode and c) working electrode (glassy carbon electrode).

Electrode (CE) (Figure 3.13). They are immersed in a solution containing an analyte and a background electrolyte. The electrochemical information about the analyte in this study was extracted from the current flowing through the WE at a potential.

During the data collection, the potential (E) of the WE is monitored with respect to constant potential of RE. The RE contains a chemical system of constant composition and known potential. Typically the $Ag|AgCl|KCl_{3M}$ system is selected as the RE because it is used for its reproducibility, reliability and convenience. In an electrochemical workstation, the voltage changes ensure that almost no current flows through the RE. This protects the RE from the negative effects of large currents which can cause changes in composition and thus in the potential of RE. The presence of the CE also minimizes the amount of uncompensated ohmic drop ($\Delta E_{ohmic} = IR$) between the WE and RE which is proportional to the resistance (R) and current (I) flowing through between the WE and RE.

Figure 3.14 shows the experimental setup (Autolab) that was used for cyclic voltammetry, amperometry and impedance analysis. The Potentiostat/Galvanostat (Autolab; Netherlands) device coupled with a Pentium IV personal computer which is installed with the General Purpose Electrochemical System (GPES) and Frequency Response Analysis (FRA) software to provide the potential and monitor the currents of the electrodes and measuring the electrochemical electron spectroscopy (EIS).

3.7 CYCLIC VOLTAMMETRY

Cyclic voltammetric studies were performed by scanning the potential of working electrode in cyclic manner by applying a potential in one direction and switching it in reverse direction. The current flow can be measured through the working electrode.

The plot of the current vs. potential is called as cyclic voltammogram. The cyclic voltammogram is recorded in stationary solutions that contain an excess of ions of a

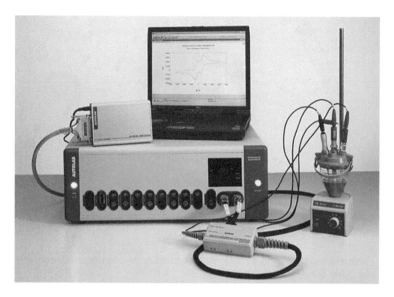

Figure 3.14 Electrochemical setup (Autolab, Metrohm).

background electrolyte over the analyte in order to suppress the migration as a mode of the analyte transport to electrode surface. Under such conditions, the analyte is transported by diffusion to the electrode surface which simplifies the mathematical description of the cyclic voltammograms.

To study the electrochemical behavior of fabricated electrodes, the cyclic voltammetric response of analyte is investigated by successively adding various concentrations of analyte into the background electrolyte. The sensitivity of electrochemical sensors was obtained by Linear Dynamic Range (LDR) and Limit of Detection (LOD).

The reproducibility and storage stability of modified electrode were examined by Relative Standard Deviation (RSD). To study the properties of modified electrode surface, Electrochemical Impedance Spectroscopy (EIS) was employed. All EIS experiments of this book were measured conducted using three electrodes (WE, CE and RF) setup with respect to Open Circuit Potential (OCP).

In analytical chemistry, the detection limit, lower limit of detection, or LOD (limit of detection), is the lowest quantity of a substance that can be distinguished from the absence of that substance (a blank value) within a stated confidence limit (generally 1%). The detection limit is estimated from the mean of the blank, the standard deviation of the blank and some confidence factor. Another consideration that affects the detection limit is the accuracy of the model used to predict concentration from the raw analytical signal.

There are a number of different "detection limits" that are commonly used. These include the Instrument Detection Limit (IDL), the Method Detection Limit (MDL), the Practical Quantification Limit (PQL), and the Limit of Quantification (LOQ). Even when the same terminology is used, there can be differences in the LOD according to nuances of what definition is used and what type of noise contributes to the measurement and calibration.

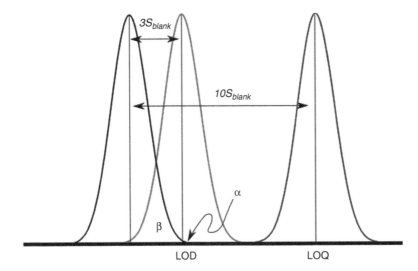

Figure 3.15 Illustration of the concept of detection limit and quantitation limit by showing the theoretical normal distributions associated with blank, detection limit, and quantification limit level samples.

Figure 3.15 illustrates the relationship between the blank, the limit of detection (LOD), and the Limit of Quantification (LOQ) by showing the probability density function for normally distributed measurements at the blank, at the LOD defined as 3 ∗ standard deviation of the blank, and at the LOQ defined as 10 ∗ standard deviation of the blank. For a signal at the LOD, the alpha error (probability of false positive) is small (1%). However, the beta error (probability of a false negative) is 50% for a sample that has a concentration at the LOD (red line). This means a sample could contain an impurity at the LOD, but there is a 50% chance that a measurement would give a result less than the LOD. At the LOQ (blue line), there is minimal chance of a false negative.

Chapter 4

Electrocatalytic detection of glucose using Cobalt (II) ions immobilized ZnO/MWCNT/PCL nanocomposite film

4.1 INTRODUCTION

Zinc Oxide (ZnO) is a unique and key inorganic material that has attracted extensive research due to its characteristic features and novel applications in a wide range of areas in science and technology. It has multiple properties such as semiconducting, piezoelectric, pyroelectric, catalytic and optoelectronics (Fan & Lu, 2005; Wang, 2004). In addition, the optical properties of ZnO play a very important role in optoelectronic, catalytic and photochemical properties (Sharma *et al.*, 2011).

In the recent years, material scientists all over the world have used different preparative techniques like Chemical Vapor Deposition (CVD), electro-deposition (ED), hydrothermal, sol-gel, vapor-liquid-solid, pulsed layer deposition, layer-by-layer method and thermal decomposition for the preparation of ZnO nanoparticles. The varied morphology, dimensionality and controlled growth of ZnO were stimulated because of strong dependency of its properties on the size, architecture and the orientation ratio of different morphologies (Cheng *et al.*, 2007; Cho *et al.*, 2008; Eftekhari *et al.*, 2006; Jin *et al.*, 2007; Kim *et al.*, 2007; Liu & Zeng, 2003; Mohanty *et al.*, 2007). The morphology and size of the nanoparticles can be controlled by using a suitable polymerizing agent such as polyethylene glycol (PEG) (Cheng & Samulski, 2004), cetyltrimethyl ammonium bromide (CTAB) (Zhai *et al.*, 2008), glycine (Narang *et al.*, 2012) and poly (sodium 4-styrene-sulfonate) (PSS) (Jiaheng Wang *et al.*, 2010) in sol-gel process. Gelatin is one type of denaturation products of collagen and it consists of a single chain of amino acids. It is soluble in warm water and becomes a gel at a concentration higher than 1wt% (Belton *et al.*, 2004; Dickerson *et al.*, 2004). A gelatin hydrogel is a three-dimensional hydrophilic polymer network which can provide a desirable water-rich buffering environment because of its attractive properties of film forming ability, biocompatibility, non-toxicity, high mechanical strength and cheapness (De Wael *et al.*, 2010).

Glucose is a major component of animal and plant carbohydrate. The quantitative determination of glucose has a very important role in biochemistry, clinical chemistry and food chemistry (Wang *et al.*, 2003; Wu *et al.*, 2004; Yildiz *et al.*, 2005). Various methods such as spectrophoto-metry, amperometry, chromatography, polarometry and capillary electrophoresis have been reported for the detection of glucose (Wu *et al.*, 2004). However, most of the current adopted methods are time consuming and costly. So, it can be expected that, they can be easily detected by using electrochemical measurements. However, their electrochemical oxidation is not so easy, as it requires a large overpotential on conventional electrodes. The most promising approach for doing so is the use of Chemically Modified Electrodes (CMEs) containing selected redox species. To date, various attempts have been made to fabricate the CMEs from nickel oxide (Sattarahmady *et al.*, 2010; Shamsipur *et al.*, 2010), ruthenium oxide (Chen *et al.*, 1993; Joseph Wang & Taha, 1990), copper oxide (Zadeii *et al.*, 1991) and cobalt macrocyclics (Barrera *et al.*, 2006) have been examined for the oxidation of glucose.

In this research, a simple and low-cost technique to synthesize the ZnO nanoparticles using gelatin as organic precursor is demonstrated. Techniques of XRD and SEM were used to characterize the architecture and properties of ZnO. When the mixed solution of ZnO, MWCNT and PCL was cast on GC electrode, a porous film was formed at the electrode surface. Therefore, the ZnO/MWCNT/PCL film was readily used as an immobilizing matrix to entrap the Co(II) ions. The Co(II) ions enhanced electrocatalytic activity towards the sensitivity of glucose at the Co/ZnO/MWCNT/PCL film modified glassy carbon electrode was achieved. The results showed that the ZnO/MWCNT/PCL composite film may be an ideal supporting material for Co(II) ions.

4.2 EXPERIMENTAL PROCEDURE

In a typical synthesis, ZnO nanoparticles are synthesized using a standard sol-gel technique mentioned in (Farley *et al.*, 2004). About 11 g of zinc nitrate was dissolved in 50 ml of deionized water and then stirred for 30 min. After that, 5 g of gelatin was dissolved in 100 ml of deionized water and stirred for 30 min at 60°C to achieve a clear gelatin solution. Then, the zinc nitrate solution was added to gelatin solution and the solution was placed in a water bath. The temperature of water bath was maintained at 80°C and the stirring was continued overnight to obtain a brown color resin. This resin became hard after the temperature of the container was reduced to room temperature. The final product was put in the muffle furnace at the temperature of 500°C for 8 hours to obtain the ZnO nanoparticles.

4.2.1 Characterization of ZnO nanoparticles

XRD pattern of the synthesized ZnO-NPs is illustrated in Figure 4.1, with comparing the main peaks with the reference sample (Zak *et al.*, 2011) crystallization of almost pure ZnO-NPs will be ascertained. The TEM image of ZnO-NPs is a useful approach that one can determine the size and morphology of obtained crystals. TEM micrograph of the synthesized sample is illustrated in Figure 4.2. Figure 4.3 shows the FTIR of the ZnO-NPs prepared by the sol-gel method, in the range of 4000–400 cm^{-1}.

The room temperature UV-Vis absorption spectra of ZnO-NPs are shown in Figure 4.4. The ZnO-NPs were dispersed in deionized water with concentration of 0.1% wt and then the solution was used to perform the UV-Vis measurement. The spectrum shows an absorption peak of ZnO at wavelength of 370 nm which can

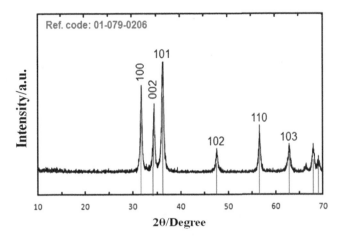

Figure 4.1 XRD pattern of ZnO-NPs.

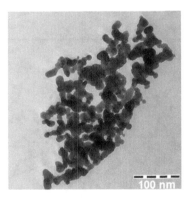

Figure 4.2 TEM image of ZnO-NPs.

be assigned to the intrinsic bandgap absorption of ZnO due to the electron transitions from the valence band to the conduction band (Yu *et al.*, 2006).

The band gap energy (*E*) was calculated as per the literature report using the following equation (Hoffmann *et al.*, 1995):

$$\text{Band gap energy } (E) = \frac{hc}{\lambda} \tag{4.1}$$

where *h* is the Plank's constant, 6.625×10^{-34} Js, *c* is the speed of light, 3.0×10^8 m/s; λ is the wavelength (m). According to this equation, the band gap of synthesized ZnO-NPs is 3.3 eV.

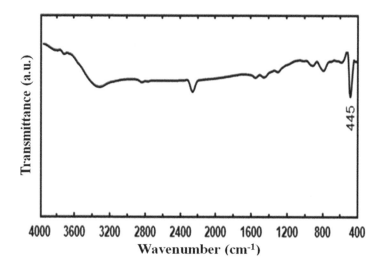

Figure 4.3 FTIR spectra of the ZnO-NPs.

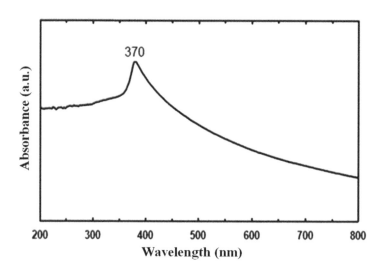

Figure 4.4 UV-Vis absorbance spectrum of the ZnO-NPs.

4.3 RESULTS AND DISCUSSION

4.3.1 Electrochemical measurement of Co/ZnO/MWCNT/PCL/GCE electrode

Chemical Modified electrode of Co/ZnO/MWCNT/PCL/GCE fabricated according to the structure which mentioned in 3.5.1. Figure 4.5A shows the cyclic voltammogram (CVs) of Co/PCL/GCE, Co/ZnO/PCL/GCE, Co/MWCNT/PCL/GCE and

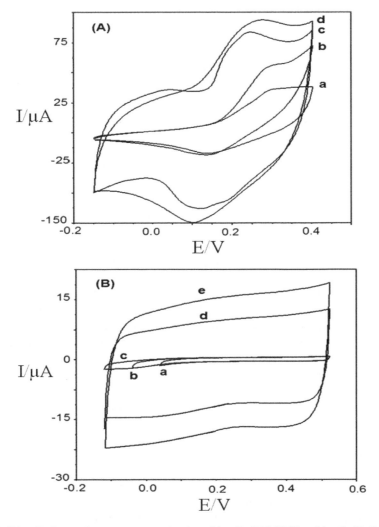

Figure 4.5 (A) Cyclic voltammograms of the (a) Co/PCL/GCE, (b) Co/ZnO/PCL/GCE, (c) Co/MWCNT/PCL/GCE and (d) Co/ZnO/MWCNT/PCL/GCE in 0.1 M NaOH solution at a scan rate of 50 mV s^{-1}. (B) Cyclic voltammograms of the (a) GCE, (b) PCL/GCE, (c) ZnO/PCL/GCE, (d) MWCNT/PCL/GCE, (e) ZnO/MWCNT/PCL/GCE.

Co/ZnO/MWCNT/PCL/GCE in 0.1 M NaOH solution at a scan rate of 50 mV s^{-1}. The defined peak currents occurred at the Co/PCL/GCE electrode (Figure 4.5A, curve a) was due to the good permeability Co (II) in the PCL film. After modification of glassy carbon electrode with Co/ZnO/PCL (Figure 4.5A, curve b) film, the peak current was increased, indicating that the ZnO nanoparticles can improve the electroactivity on the surface of the electrode.

Comparing this with the Co/MWCNT/PCL/GCE (Figure 4.5A, curve c), the enhanced electron transfer of the entrapped Co in Co/MWCNT/PCL composite film

Table 4.1 Anodic and cathodic potential, the half-wave potential and the peak-to-peak separation potential for redox couple in cyclic voltammogram of Figure 4.5A.

Electrode	I_{pc} (μA)	I_{pa} (μA)	ΔE_p (V)	$E_{1/2}$ (V)	E_{pc} (V)	E_{pa} (V)
Co/PCL/GCE	−21	31	0.140	0.200	0.130	0.270
Co/ZnO/PCL/GCE	−22	53	0.133	0.197	0.130	0.264
Co/MWCNT/PCL/GCE	−65	78	0.092	0.154	0.108	0.200
Co/ZnO/MWCNT/PCL/GCE	−75	87	0.100	0.160	0.110	0.210

indicated that, the MWCNT played a key role in facilitating the electron transfer between Co(II) ions and glassy carbon electrode. When Co/ZnO/MWCNT/PCL film was modified on the glassy carbon electrode surface, a couple of quasi-reversible and well-defined peaks with the apparent formal peak potential ($E°$) of 160 mV and peak-to-peak separation (ΔE_p) of 100 mV was observed at 50 mV s^{-1} for Co/ZnO/MWCNT/PCL/GCE and the peak current was distinctly increased due to the formation of conduction pathway in the ZnO/MWCNT/PCL composite film (Figure 4.5A, curve d). The cyclic voltammograms of GCE (Figure 4.5B, curve a), PCL/GCE (Figure 4.5B, curve b), ZnO/PCL/GCE (Figure 4.5B, curve c), MWCNT/PCL/GCE (Figure 4.5B, curve d) and ZnO/MWCNT/PCL/GCE (Figure 4.5B, curve e) in 0.1 M NaOH solution showed no anodic or cathodic peaks.

The effect of scan rates on the response of Co/ZnO/MWCNT/PCL/GCE is shown in Figure 4.6A. With increasing the scan rate, both reduction and oxidation peak currents (I_p) were increasing linearly with the scan rates (Figure 4.6B), indicating the surface-controlled process.

Anodic and cathodic potential, the half-wave potential and also the peak-to-peak separation potential for redox couple in each of the modified electrode was measured according to the reference electrode potential in cyclic voltammogram of Figure 4.5A and presented in table 4.1. It is clear that Co/ZnO/MWCNT/PCL/GCE reduced the overpotential.

A plot of E_p versus log v yields straight line with slopes of $-2.3RT/\alpha nF$ and $2.3RT/(1-\alpha)nF$ for cathodic and anodic peak, respectively. As a result the α can be estimated as 0.57 from of the slope. According to this equation, $\Gamma = Q/nFA$, the surface coverage (Γ) was estimated from the integration of oxidation or reduction peaks. Where Q is the charge consumed in coulombs, obtained from integrating the anodic (or cathodic) peak area in cyclic voltammograms under the background correction. The average γ value of $(3.24 \pm 0.98) \times 10^{-10}$ mol/cm^2 was obtained.

4.3.2 Electrochemical impedance spectroscopy

Electrochemical impedance spectroscopy was employed to further investigate the impedance changes of the electrode surface in the modified process Figure 4.7 illustrates the EIS profiles of different modified electrodes in which the Nyquist plots are shown with the real part (Z_I) on the X-axis and the imaginary part (Z_R) on the Y-axis. The spectra showed well-defined semicircles at high frequencies followed by straight line at 45° to the real impedance Z_R at low frequencies. The semicircular portion at higher

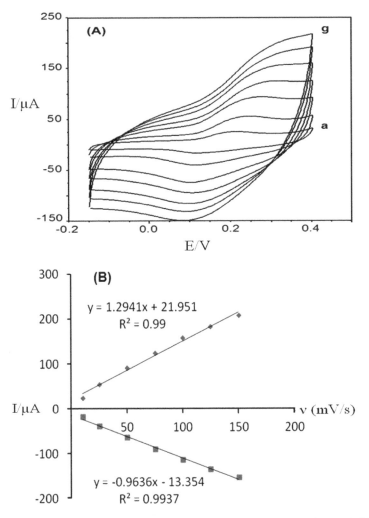

Figure 4.6 (A) Cyclic voltammograms of Co/ZnO/MWCNT/PCL/GCE in 0.1 M NaOH solution at different scan rates. The scan rates are: (a) 10, (b) 25, (c) 50, (d) 75, (e) 100, (f) 120 and (g) 150 mV s^{-1}, respectively. (B) The plot of cathodic and anodic peak currents vs scan rates.

frequencies corresponds to electron-transfer-limited process with its diameter equal to the electron transfer resistance (R_{ct}), which controls the electron transfer kinetics of the redox couple at the electrode interface.

Meanwhile, the linear part at lower frequencies corresponds to the diffusion process. The largest R_{ct} on the bare electrode (curve a) indicates the slow electron transfer rate between the bare glassy carbon electrode and $[Fe(CN)_6]^{3-/4-}$ couples. The Nyquist diameter of the PCL modified glassy carbon electrode (curve b) is smaller than that of the bare glassy carbon electrode, which suggests that the PCL film coated on the glassy carbon electrode is a porous material with a large surface area and increases the electron transfer of the redox probe.

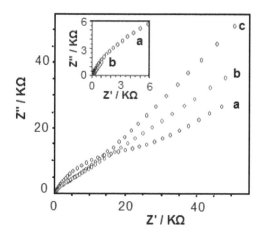

Figure 4.7 Impendence plots of (a) bare GCE, (b) PCL/GCE and (c) ZnO/PCL/GCE in the presence of 1.0 mM $[Fe(CN)_6]^{3-/4-}$ containing 0.1 M KCl as supporting electrolyte. Inset: (a) EIS of (a) MWCNT/PCL/GCE and (b) ZnO/MWCNT/PCL/GCE in the same condition.

However, when the ZnO/PCL was cast on the surface of the electrode (curve c), the R_{ct} was smaller than that of PCL/GCE, which was attributed to the good conductivity of ZnO decreased the resistance to the electrochemical reaction of $[Fe(CN)_6]^{3-/4-}$ couples. Comparing with the ZnO/PCL film modified glassy carbon electrode, an obvious decrease in the interfacial resistance observed in the MWCNT/PCL film modified glassy carbon electrode (Inset: curve a), which indicates that MWCNT was immobilized successfully into the film. After modifying GCE with ZnO/MWCNT/PCL, a lowest interfacial resistance was observed (Inset: curve b). These data showed that the ZnO, MWCNT and PCL film have been successfully attached to the electrode surface and formed a tunable kinetic barrier.

4.3.3 Electrocatalytic properties of Co/ZnO/MWCNT/PCL/GCE electrode

In order to investigate the activity of Co(II) at glassy carbon electrode modified with ZnO/MWCNT/PCL composite film, its response to the oxidation of glucose was studied. Figure 4.8 shows the cyclic voltammogram of Co/ZnO/MWCNT/PCL/GCE in the presence of glucose. When glucose was added to 0.1 M NaOH solution, the oxidation peak current increased obviously accompanied by the decrease in reduction peak current which indicates a typical electrocatalytic oxidation process and also which may be due to the reaction of Co(II) with glucose.

Furthermore, the oxidation peak current increases with increasing concentration of glucose. These results further confirmed that, the ZnO/MWCNT/PCL film provided a friendly platform for the immobilization of Co(II) and the electrocatalysis to glucose. The electrocatalytic process can be described as follows:

$$Co(II) \rightarrow Co(III) + e^-$$

$$Co(III) + glucose \rightarrow Co(II) + products$$

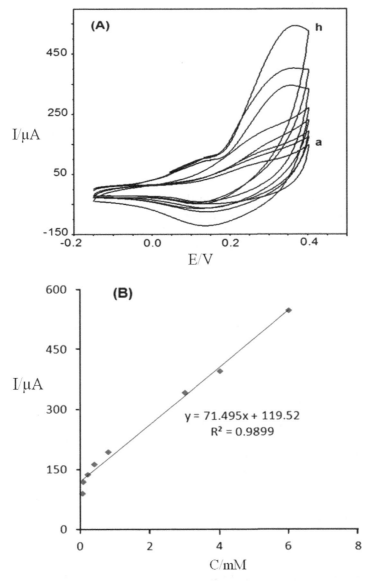

Figure 4.8 Cyclic voltammograms of Co/ZnO/MWCNT/PCL/GCE in the presence of (a) 0.05,
(b) 0.08, (c) 0.20, (d) 0.40, (e) 0.80, (f) 3.00, (g) 4.00 and (h) 6.00 mM of glucose in 0.1 M NaOH
solution at scan rate 50 mV s⁻¹. (B) Plot of anodic peak current vs. glucose concentration.

Co(III) acts as strong oxidants, reacting with the glucose molecules to yield a radical. Further reaction of radical with additional surface sites results in the formation of
product. As reported in the literature, gluconolactone and gluconic acid were detected
as the main products of oxidation of glucose (Tominaga *et al.*, 2005; Zhao *et al.*,
2007). In addition to this, formate and oxalate were reported as oxidation products
(Mho & Johnson, 2001).

Table 4.2 The analytical recoveries of glucose solutions added to 0.1 M NaOH solution, suggesting better accuracy of the method.

Modifier	Sensitivity $(\mu A\ mM^{-1}cm^{-2})$	LDR (μM)	LOD (μM)	References
Pt-Pb nanowire	11.25	0–11000	8	(Bai et al., 2008)
OMC[a]	10.81	500–2500	20	(Chen et al., 2010)
Ni/MWCNT	–	200–12000	160	(Shamsipur et al., 2010)
CNT-NiCo	66.15	10–12120	5	(Arvinte et al., 2011)
Co/ZnO/MWCNT/PCL/GCE	567.56	50–6000	16	This work

[a] Ordered mesoporous carbon

The cyclic voltammetric response of the Co/ZnO/MWCNT/PCL/GCE to glucose is illustrated in Figure 4.8. The voltammetric response showed a linear relation with glucose concentration from 5.0×10^{-5} to 6.0×10^{-3} M with the detection limit of 1.6×10^{-5} M (S/N = 3) comparable with the results reported for analytical determination of glucose at the surface of different modified electrodes (Arvinte et al., 2011; Bai et al., 2008; Chen et al., 2010; Shamsipur et al., 2010) (see Table 4.2). The results show that linear dynamic range (LDR) and limit of detection (LOD) in comparison to other researchers is not very good but the sensitivity of the method is excellent. The oxidation peak potential of glucose shifts to the less positive potential at Co/ZnO/MWCNT/PCL/GCE is more than that at different modified electrodes reported in Table 4.2.

4.3.4 Interference study

Possible interference for the detection of glucose at Co/ZnO/MWCNT/PCL/GCE film was investigated by the addition of various compounds such as sorbitol, sucrose, fructose, glutathione, cysteine, tryptophan, glycine, alanine, ascorbic acid, dopamine, uric acid, penicillamine, methionine, lysine and hydrogen peroxide to 0.1 M NaOH solution in the presence of 1.0 mM glucose. The carbohydrates such as sorbitol, fructose and sucrose showed interference to glucose determination, which found to react through a similar route, was observed with glucose, while the other compounds did not show any interference.

4.3.5 Stability and reproducibility of Co/ZnO/MWCNT/PCL/GCE electrode

The stability of the Co/ZnO/MWCNT/PCL/GCE was investigated. The sensor could retain the electrochemistry of immobilized Co(II) at constant current values upon the continuous CV sweep over the potential range from 0.15 V to 0.4 V at $50\,mV\,s^{-1}$, suggesting that the Co(II) can tightly adsorb on the surface of ZnO/MWCNT/PCL-modified glassy carbon electrode. When stored for over 2 weeks, the sensor retained 90% of the initial sensitivity to glucose. The relative standard deviation (RSD) of the sensor showed a 4.7% for 15 successive determinations in 1.0 mM glucose solution, indicating that it is a good precision.

Table 4.3 Assay of glucose in human blood serum samples and recovery of glucose in 0.1 M NaOH solution spiked with different concentrations.

Sample	Spiked (mM)	Biochemical analyzer in a local hospital (mM)	Found (mM)	Mean recovery (n = 3)
Blood serum	–	4.70	4.58	–
	–	4.57	4.46	–
Glucose	0.15	–	0.16	106 ± 3.4
	0.22	–	0.21	95 ± 4.1
	0.31	–	0.30	97 ± 3.9

4.3.6   Determination of glucose in human blood serum

The response of modified sensor to the glucose in human blood serum was investigated. The serum sample obtained from hospitalized patient and it was analyzed. The results were matched with the referenced value obtained from the automated standard calorimetric technique (D'Eramo et al., 1999) in the hospital. Table 4.3 shows analytical recoveries of the glucose solutions added to 0.1 M NaOH solution, suggesting the good accuracy of the method.

4.4   CHAPTER SUMMARY

In this work, a gel-assisted technique for the synthesis of ZnO nanoparticles was reported. The morphological features, crystallization and purity of synthesized ZnO were characterized by using XRD, FTIR, UV-Vis and TEM techniques. Due to the large surface area and high conductivity of ZnO, a novel compatible ZnO/MWCNT/PCL composite film has been successfully developed for the immobilization of Co(II) ions. This kind of composite film provided a favorable microenvironment around Co to retain its activity. Moreover, the sensor based on the Co/ZnO/MWCNT/PCL composite film modified electrodes exhibited good electrocatalytic response to the oxidation of glucose, good reproducibility and stability. Therefore, this ZnO/MWCNT/PCL composite film could offer a new promising platform for further study of direct electrochemistry of redox proteins and the development of biosensors.

Chapter 5

Sensitive detection of *L*-Tryptophan using gelatin stabilized anatase titanium dioxide nanoparticles

5.1 INTRODUCTION

Titanium dioxide (TiO_2) has proven to be a promising n-type semiconductor material because of its wide band gap (3.2 eV) under UV light (Zaleska, 2008). In addition to this, it has a high physical and chemical stability and also a high refractive index. This makes the material widely used by material scientists (Abbas *et al.*, 2011; Xie *et al.*, 2012). Several semiconductor oxides such as ZnO, TiO_2, NiO, CuO, Al_2O_3, Fe_2O_3, SnO_2, ZrO_2, and WO_3 have received considerable attention over the last few years due to their distinctive optical and electronic properties, but TiO_2 can also be used in several other domains such as photocatalysts, solar cells, sensors and bactericidal action (Chaturvedi *et al.*, 2012; Gómez *et al.*, 2003; Grieshaber *et al.*, 2008). Recently, there is a considerable interest in using TiO_2 nanoparticles as a modifier since they have a high surface area, good biocompatibility and relatively good conductivity (Chaturvedi *et al.*, 2012; Liu *et al.*, 2012).

There are several methods for synthesizing the TiO_2 nanoparticles. Some researchers have recommended using the microemulsion method due to its short processing time (Ki Do Kim *et al.*, 2003). On the other hand, by using physical vapor deposition (PVD), materials are condensed after evaporation to solid form (Prakash *et al.*, 2012). Other methods like hydrolysis (Manjari Lal, 1998) and hydrothermal (Campbell *et al.*, 1992) have been frequently used. In the sol-gel method, materials undergo hydrolysis and polycondensation to form a sol; a gel forms after aging and eventually becomes solid after drying. It can be a simple method as it requires low temperature and controllable final product properties (Rao *et al.*, 1996; Tabata & Ikada, 1998; Vieira & Pawlicka, 2010).

Gelatin is a protein and it can be derived from collagen in tissues by heat denaturation and it has a three-chain helical structure in which individual helical chains are stranded in a super-helix about the common molecular axis (Kawanishi *et al.*, 1990). Gelatin contains positively and negatively charged ions and it is stable in folded hydrophobic domains, and also non-native state minimizes its hydrophobic interactions with water (Likos, 2009). Hence, it is a suitable polymerizing agent to achieve small sized TiO_2-NPs in the sol-gel method.

L-Tryptophan (*L*-Trp) is an essential amino acid for human and herbivores bodies because it has the precursor of hormone, neurotransmitter serotonin, neurohormone melatonin and other relevant biomolecules (Liu *et al.*, 2011). It is also a vital component of proteins that could establish and maintain a positive nitrogen balance (Huang *et al.*, 2009). Owing to the insufficient presence in vegetables, *L*-Trp has been commonly added to dietary, food products and pharmaceutical formulations. However, an overdose of *L*-Trp may produce a toxic metabolite in the brain which could cause hallucinations and delusions (Walter Kochen (Editor), 1994). Thus, it is necessary to develop a simple, accurate, rapid and inexpensive method for the determination of *L*-Trp in food processing, biological fluids, pharmaceuticals and clinical analysis (Fiorucci & Cavalheiro, 2002). The electrochemical methods were employed extensively for the elegant and sensitive properties such as selectivity, simplicity and reproducibility for the determination of *L*-Trp (Fang *et al.*, 2007; Frith & Limson, 2009; Gholivand *et al.*, 2011; Goyal *et al.*, 2011; Mao *et al.*, 2012; Özcan & Şahin, 2012; Safavi *et al.*, 2006; Tang *et al.*, 2010).

As per my knowledge, no work has been done on the synthesis of TiO_2-NPs via gelatin. The present investigation is focused on the series of practical procedures for the sol-gel method to introduce the synthesis of anatase TiO_2-NPs with and without gelatin as a stabilizing agent. Furthermore, the surface morphology and size of the anatase TiO_2-NPs were identified and analyzed by using different methods. It was found that, gelatin acts as a polymerizing agent that helps in controlling nanoparticles' size and dispersion due to expansion during calcinations. Also, we have discussed in detail the preparation of TiO_2-NPs and carbon nanotubes composite thin film and modified electrode to investigate the electrochemical behavior of *L*-Trp using the voltammetric method. The modified electrode provides a suitable and effective method for the determination of *L*-Trp in pharmaceutical and clinical preparations.

5.2 EXPERIMENTAL PROCEDURE

5.2.1 Synthesis of anatase TiO_2 nanoparticles

The precursor solution of TiO_2-NPs without gelatin was prepared by dissolving titanium (IV) isopropoxide in glacial acetic acid and deionised water with the molar ratio of 1:10:200. Glacial acetic acid acts as a catalyst that can prevent the titanium (IV) isopropoxide from the nucleophilic's attacks of water (Sanchez *et al.*, 1988). The solution was stirred for several hours and dried at about 80°C overnight. The dried gel was ground to fine particles and calcined in a muffle furnace at 500°C for 5 hours. For the synthesis of TiO_2-NPs with gelatin, the precursor solution was prepared similar to the above mentioned procedure. Gelatin solution was made by dissolving 3.0 g of gelatin in 100 ml of deionized water and stirred for 30 min at 60°C to achieve a clear gelatin solution. Then, the gelatin solution was added into the precursor solution. The resulted solution was continued to be heated and stirred at about 80°C for 5 hours. The solution was then dried in an oven at 80°C overnight. The dried gel was ground and calcined in a muffle furnace at 500°C for 5 hours.

5.2.2 Fabrication of TiO_2-MWCNT/GCE

Appropriate amount of TiO_2 nanoparticles and MWCNTs were dispersed in dimethyl formamide (DMF) solution. The mass ratio of TiO_2: MWCNT was 1:5. The mixture was ultrasonicated for 20 min. Finally, a highly dispersed black colloidal solution was formed. Prior to use, the GC electrode (diameter 2 mm) was first polished with alumina (0.05 μm) slurry and it was ultrasonically cleaned with ethanol followed by double distilled water. Then, it was dried at room temperature.

5.3   RESULTS AND DISCUSSION

5.3.1   Thermogravimetric analysis

Figure 5.1 (a and b) shows the TGA curves of TiO_2-NPs with and without gelatin, respectively. TiO_2 samples of about 12.32 mg were dynamically heated from 50 to 900°C at 10°C/min. According to the curve (b) in Figure 5.1, the TiO_2-NPs without gelatin experienced two pronounced mass loss steps in the temperature ranges 30°C–110°C and 110°C–350°C, respectively. The first weight loss happened at about 105°C due to the evaporation of surface adsorbed water. The second mass losses occurred at 110–350°C due to the volatilization and combustion of organic species such as CH_3COOH. There were no associated signals with these latter thermal events in TGA curve confirming that the crystallization and phase transition events. The TiO_2-NPs which were synthesized by gelatin experienced three main step downs through the TGA test (curve a). The first weight loss between 30 and 110°C was an initial loss of water.

The second step of the weight loss (110 to 350°C) was related to both, that is the decomposition of chemically bonded organic groups and the formation of pyrochlore phases. The last weight-loss step was from 350 to 460°C and was attributed to the decomposition of pyrochlore phases and the formation of TiO_2-NPs pure phases. No

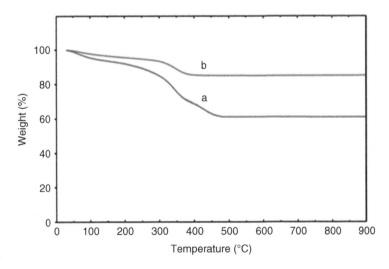

Figure 5.1   TGA curves of TiO_2-NPs synthesized with (curve a) and without (curve b) gelatin from 50°C to 900°C.

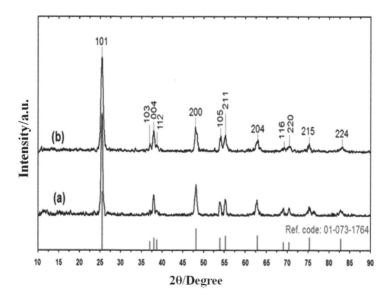

Figure 5.2 XRD patterns of TiO$_2$-NPs synthesized with (curve a) and without (curve b) gelatin.

weight loss was observed between 460 and 900°C, which indicates that the formation of nanocrystalline TiO$_2$ particles as the decomposition product.

5.3.2 X-ray diffraction to study TiO$_2$ nanoparticles and poly dispersity index

Figure 5.2 (a and b) shows the XRD patterns of the TiO$_2$-NPs prepared with and without gelatin. The XRD peaks in wide angle range of 2θ ($10° < 2\theta < 90°$) ascertained that the peaks were in 25.36°, 37.05°, 37.90°, 38.66°, 48.15°, 54.05°, 55.20°, 62.86°, 68.97°, 70.47°, 75.30° and 82.92° could be attributed to the (101), (103), (004), (112), (200), (105),(211), (204),(116), (220), (215) and (224) (*hkl* values) crystalline structures of the anatase structure of TiO$_2$-NPs, respectively (*Anatase* XRD Ref. No 01-073-1764).

By using Scherrer's formula, the crystallite size can be determined. The (101) plane was chosen to calculate the crystallite size (either plane can be used for this purpose). The average crystallite size for the synthesized TiO$_2$-NPs with and without gelatin was approximately about 12 nm to 19 nm, respectively.

The poly dispersity index (PDI) is a parameter used to determine the size distribution of the particles in a sample. It is also defined as the ratio between the average weight of the molecular weight and the number average of the molecular weight. In addition, the PDI value for monodispersed particles ranges from 0.01 to 0.7 whereas the large size distribution is above 0.7. PDI for TiO$_2$-NPs with and without gelatin as a stabilizing agent is 0.225 and 0.233, respectively. It is clear that these particles are monodispersed.

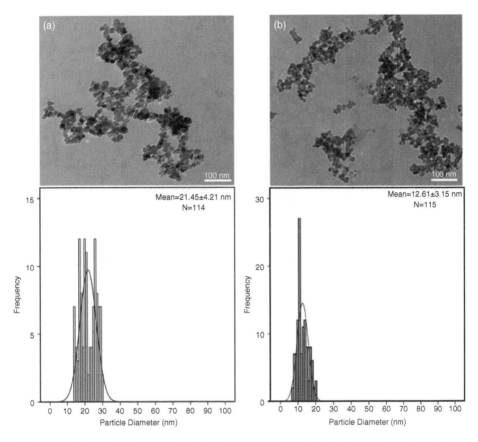

Figure 5.3 TEM images and the particle size of TiO$_2$-NPs synthesized with (curve a) and without (curve b) gelatin.

5.3.3 Transmission of electron microscopy

The TEM images of TiO$_2$-NPs prepared with and without gelatin are shown in Figure 5.3 (a and b) and it shows morphology and size distribution of the materials. The size histograms of the TiO$_2$-NPs are shown beside the relative TEM images. The histograms show that the main particle size of the TiO$_2$-NPs prepared with and without gelatin were about 12.61 ± 3.15 and 21.45 ± 4.21 nm, respectively. The images illustrate that the samples are close to spherical morphology.

5.3.4 Fourier transform infrared spectroscopy (FTIR)

Figure 5.4 (a and b) shows the FTIR spectra of TiO$_2$-NPs synthesized via the sol-gel method with and without gelatin in the range of 400–4000 cm^{-1} respectively. The Figure 5.4a depicts that, the synthesized TiO$_2$-NPs by gelatin shows the peaks at 468 cm^{-1} and 712 cm^{-1} are for O-Ti-O bonding (Yu *et al.*, 2003).

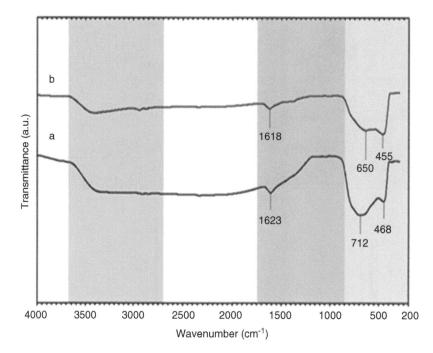

Figure 5.4   FTIR spectra of TiO$_2$-NPs synthesized with (curve a) and without (curve b) gelatin.

The band centred at 1623 cm^{-1} is a characteristic peak of δ-H$_2$O bending (Karakitsou & Verykios, 1993). Figure 5.4b shows the synthesized TiO$_2$-NPs without gelatin, the peaks at 455 cm^{-1} and 650 cm^{-1} are for O-Ti-O bonding. The band centred at 1618 cm^{-1} is a characteristic peak of δ-H$_2$O bending.

5.3.5   UV-Vis absorption spectroscopy (UV-Vis)

UV-Vis absorption spectra of TiO$_2$-NPs synthesized with and without gelatin between 200 nm and 700 nm are shown in Figure 5.5 (a and b). Synthesized TiO$_2$-NPs without gelatin have peak at 380 nm and synthesized TiO$_2$-NPs with gelatin have peak at 376 nm. It is clear that as the size of particles decreases peaks become sharp. The blue shift is ascribed to decrease in crystallite size.

The band gap energy (*E*) was calculated as per the literature report using the following equation (Hoffmann *et al.*, 1995):

$$\text{Band gap energy } (E) = \frac{hc}{\lambda} \tag{5.1}$$

where *h* is the Plank's constant, 6.625×10^{-34} Js, *c* is the speed of light, 3.0×10^8 m/s; λ is the wavelength (m). According to this equation, the band gap of synthesized TiO$_2$ NPs without gelatin is 3.2 eV and the band gap of synthesized TiO$_2$-NPs using gelatin

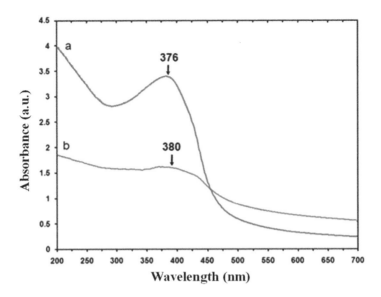

Figure 5.5 UV-Vis spectra of TiO$_2$-NPs synthesized with (curve a) and without (curve b) gelatin.

is 3.3 eV. These results confirmed that when the size is smaller, the band gap will be bigger.

5.4 ELECTROCATALYTIC PROPERTIES OF TiO$_2$-MWCNT/GCE ELECTRODE

In order to investigate the electrocatalytic activity of TiO$_2$ at GCE modified with TiO$_2$-MWCNT composite film, its response to the oxidation of *L*-Trp was studied. Figure 5.6 shows the electrooxidation of 0.1 mM *L*-Trp in 0.1 M phosphate buffer solution (PBS, pH 7.00) at bare GCE (curve a) MWCNT/GCE (curve b) and TiO$_2$-MWCNT/GCE (curve c) electrodes by cyclic voltammetry. As shown in the cyclic voltammograms, the oxidation peak current increased obviously at the surface TiO$_2$-MWCNT/GCE, indicating that the presence of TiO$_2$-NPs at the surface electrode. This is due to its high surface area and relatively good conductivity that facile the electrooxidation of *L*-Trp and decreases its anodic potential towards less positive potential. Therefore, the electro-oxidation of *L*-Trp leads to an increase in the anodic current.

5.5 DETERMINATION OF *L*-TRP

The *L*-Trp electro-oxidation was explored for amperometric responses of different concentrations of *L*-Trp at TiO$_2$-MWCNT/GCE with an applied potential of 700 mV in 0.1 M PBS (pH 7.00) (Figure 5.7). We used the current value to plot with the concentration of *L*-Trp (Inset). There was a linear relation of current with concentration

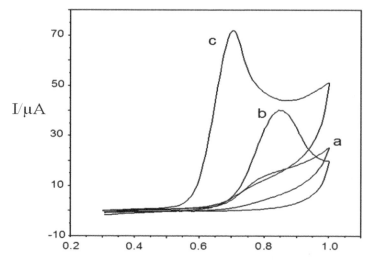

Figure 5.6 Cyclic voltammograms of (a) GCE, (b) MWCNT/GCE, (c) TiO$_2$-MWCNT/GCE in 0.1 M PBS solution (pH 7.00) and 0.1 M KCl as supporting electrolyte at scan rate of 50 mV s^{-1}.

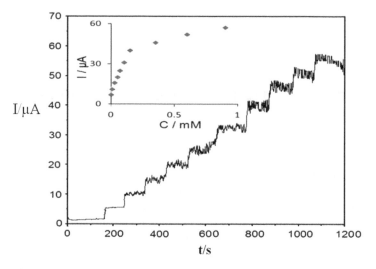

Figure 5.7 Current–time curve of TiO$_2$-MWCNT/GCE with successive addition of Trp to a stirred 0.1 M PBS (pH 7.00). The inset is the calibration curve.

of *L*-Trp between 1.0×10^{-6} to 1.5×10^{-4} M. The detection limit was 5.2×10^{-7} M of *L*-Trp with the signal to noise ratio was around 3.

The operational stability of TiO$_2$-MWCNT/GCE was tested once in a week by cyclic the voltammetric method. The response to 0.1 mM *L*-Trp decreased less than 14% after 2 weeks, so it has good stability. The fabrication reproducibility of five electrodes independently was made and it was showed a RSD. of 7.1% for detecting 0.1 mM *L*-Trp.

5.6   CHAPTER SUMMARY

Anatase TiO_2-NPs were successfully synthesized by simple sol-gel method with and without gelatin. XRD and FTIR results clearly show that synthesized TiO_2-NPs exhibited anatase structure. The TEM and XRD results indicated that using gelatin as a stabilizer decreased the particle size of the TiO_2-NPs. The particle size of the TiO_2-NPs prepared without gelatin were about 21 nm, while that of TiO_2-NPs synthesized with gelatin were about 13 nm. This result confirms that the gelatin is a good stabilizer and polymerizing agent for the synthesis of TiO_2-NPs by the sol–gel method. It was also demonstrated that TiO_2/MWCNT could yield a new platform to facilitate electron transfer. Stable TiO_2/MWCNT/GCE can be used to determine *L*-Trp. This demonstration indicates that, this is a new way of synthesizing the gelatin based TiO_2/MWCNT composite sensors, catalytic bioreactors and biomedical devices, etc.

Chapter 6

Electrocatalytic detection of carbohydrates by using Ni(II)/PAN/MWCNT nanocomposite thin films

6.1 INTRODUCTION

Direct electrochemical oxidation of sugars is of great importance for several points of view ranging from biomedical (blood sugar analysis) and fuel cell applications to ecological approaches (wastewater treatment) (Meng et al., 2009; Newman & Turner, 2005; Shoji & Freund, 2001; Wu et al., 2007; You et al., 2003). The detection of glucose level in the blood is a challenging task from diagnostic and therapy of diabetics. The electrochemical biosensors have been applied successfully for the determination of glucose (Fu et al., 2009; Gopalan et al., 2009; Sheng et al., 2009; Tasviri et al., 2011; Wu et al., 2009; Yang et al., 2011; Hongfang Zhang et al., 2011). Some of the electrodes like platinum (de Mele et al., 1983), copper (Hampson et al., 1972; Luo & Baldwin, 1995; Torto et al., 1999), nickel (Fleischmann et al., 1971, 1972) and gold (Matsumoto et al., 2003; Parpot et al., 2006), and also on modified surfaces such as ruthenium dioxide (Joseph Wang & Taha, 1990), nickel oxide (Berchmans et al., 1995; Vidotti et al., 2009), cobalt oxide (Buratti et al., 2008), alloy (Yeo & Johnson, 2001) and metallic complexes such as cobalt phthalocyanine (Santos & Baldwin, 1987) have been explored to investigate the direct electrochemical oxidation of sugars in alkaline medium. Among the various possible electro-catalysts, nickel hydroxide has attracted much attention specifically as a fuel cell catalyst, secondary batteries and electro-catalyst for organic synthesis (Burda et al., 2005; Daniel & Astruc, 2003). Its unique electro-catalytic effect arises from the unpaired 'd' electrons and vacant 'd' orbitals associated with the oxidized form of nickel oxyhydroxide that are readily available to bind any adsorbed species (Lo & Hwang, 1995).

It is also well known that nickel and nickel hydroxide exhibits an excellent electro-catalytic behavior for alkaline medium (Hutton et al., 2010). An inert porous material with a large specific surface area and high permeability could be a promising candidate for efficient catalyst carriers. Porous polymer materials having polar groups can satisfy these requirements. Different porous polymeric materials have found wide application as immobilization matrix (Lim et al., 2005; Nau & Nieman, 1979; Reddy & Vadgama, 1997; Rubtsova et al., 1998; Ying et al., 2002). In this manner, a necessary condition is that the size of micropore should be a little larger than the size of an immobilized particle. Polyacrylonitrile (PAN) has been successfully applied as membrane materials in the fields of dialysis (Lin & Yang, 2003), ultrafiltration (Nie et al., 2004), enzyme-immobilization (Shan et al., 2006) and evaporation (Bhat & Pangarkar, 2000).

Carbon nanotubes can be functionalized covalently or non-covalently with various polymers. Non-covalent interactions such as π-π interaction, π-cation interaction and ionic interaction between MWCNTs and polymers enables the absorption of polymers onto the MWCNT surfaces (Kim & Jo, 2007; Lee *et al.*, 2007; Park *et al.*, 2008; Wang *et al.*, 2006).

The purpose of this study was to fabricate Ni/PAN-MWCNT conductive composite film. This study also aimed to conduct detailed investigations of electrocatalytic oxidation of carbohydrates on Ni/PAN-MWCNT composite film in alkaline solution. It is well known that, MWCNTs have high-surface-area and when combined with metal oxide, they can improve the performance of the final material, sensitivity and stability of the electrochemical sensors. The electrocatalytic oxidation of carbohydrates on this type of electrode is attractive because of the interest and needs in sensitive and stable sugar sensors for medical and food industry, as well as sugar–oxygen fuel cell application.

6.2 EXPERIMENTAL PROCEDURE

6.2.1 Preparation of modified electrodes

Chemical Modified electrode of Ni/MWCNT/GCE, Ni/PAN/GCE fabricated according to the structure which mentioned in 3.5.3.

6.3 RESULTS AND DISCUSSION

6.3.1 Properties of modified electrode

Figure 6.1 shows the cyclic voltammetric responses obtained at the Ni/GCE, Ni/MWCNT/GCE, Ni/PAN/GCE and Ni/PAN-MWCNT/GCE in 0.1 M NaOH solution. The anodic and cathodic peaks corresponding to Ni(III)/Ni(II) couple are observed in both the cases. Although the cathodic and anodic peak potentials are nearly the same for modified electrodes, the cathodic and anodic peak current for Ni/PAN-MWCNT/GCE increased significantly, compared to those of the Ni/GCE, Ni/MWCNT/GCE and Ni/PAN/GCE. The increase in peak current results from a large surface area of the Ni/PAN-MWCNT/GCE.

The redox potential of Ni/PAN-MWCNT/GCE is dependent on the scan rate. Figure 6.2(A) shows the cyclic voltammograms of Ni/PAN-MWCNT/GCE in 0.1 M NaOH at various scan rates. It was observed that the values of E_{pa} and E_{pc} shift slightly to the positive and negative directions, respectively, and ΔE_p increases with an increase in the scan rates. The anodic and cathodic peak currents are linearly proportional to the scan rates (Figure 6.2(B)), suggesting that the reaction is not a diffusion-controlled process but a surface-controlled one, as expected for immobilized systems (Shamsipur *et al.*, 2010).

A plot of E_p versus log ν yields straight line with slopes of $-2.3RT/\alpha nF$ and $2.3RT/(1-\alpha)nF$ for the cathodic and anodic peak, respectively, so that α can be

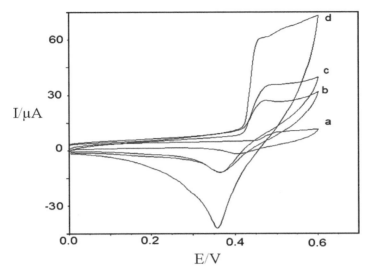

Figure 6.1 Cyclic voltammograms of the Ni/GCE (a), Ni/MWCNT/GCE (b), Ni/PAN/GCE (c) and Ni/PAN-MWCNT/GCE (d) in 0.1 M NaOH solution at a scan rate of 50 mV s^{-1}.

estimated as 0.62 from the slope of straight lines. The surface concentration of electroactive Ni on Ni/PAN-MWCNT/GCE, Γ (in mol/cm^2), can be estimated using the equation (Laviron, 1979):

$$\Gamma = Q/nFA \qquad (6.1)$$

where Q is the charge consumed in coulombs obtained from integrating the anodic (or cathodic) peak area in cyclic voltammograms under the background correction. The average Γ value of $(3.24 \pm 0.67) \times 10^{-10}$ mol/cm^2 was obtained.

Anodic and cathodic potential, the half-wave potential and also the peak-to-peak separation potential for redox couple in each of the modified electrode was measured according to the reference electrode potential in cyclic voltammogram of Figure 6.1 and presented in Table 6.1. It is clear that Ni/PAN-MWCNT/GCE reduced the overpotential.

6.3.2 Electrochemical Impedance Spectroscopy (EIS)

Electrochemical impedance spectroscopy was employed to further investigate the impedance changes of the electrode surface in the modified process. Figure 6.3 shows the EIS results of bare GCE, PAN/GCE and PAN-MWCNT/GCE in the presence of 1.0 mM $[Fe(CN)_6]^{3-/4-}$. To clearly understand the electrical properties of as-prepared electrodes/solution interfaces, the semicircle diameter equals the charge transfer resistance (R_{ct}). This resistance exhibits the electron transfer kinetics of the redox-probe at the electrode interface. As shown in Figure 6.3, there is the charge transfer resistance about 1.15×10^4 Ω for $[Fe(CN)_6]^{3-/4-}$ at the bare GCE (curve a). After modifying GCE with PAN film, the R_{ct} decreased to about 5.31×10^3 Ω (curve b), indicating

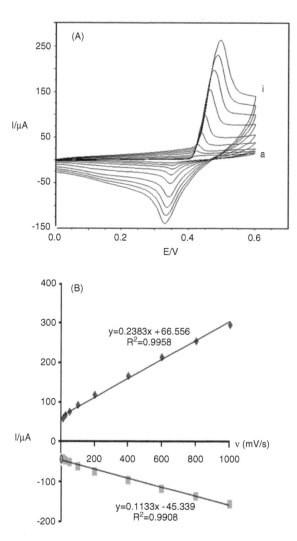

Figure 6.2 (A) Cyclic voltammograms of Ni/PAN-MWCNT/GCE in 0.1 M NaOH solution at different scan rates. The scan rates are: a) 10, b) 25, c) 50, d) 100, e) 200, f) 400, g) 600, h) 800 and i) 1000 mV s^{-1}, respectively. (B) The plot of cathodic and anodic peak currents vs. scan rates.

Table 6.1 The Anodic and cathodic potential, the half-wave potential and the peak-to-peak separation potential for redox couple of cyclic voltammogram of Figure 6.1.

Electrode	I_{pc} (μA)	I_{pa} (μA)	ΔE_p (V)	$E_{1/2}$ (V)	E_{pc} (V)	E_{pa} (V)
Ni/GCE	−4	5	0.100	0.450	0.400	0.500
Ni/CNT/GCE	13	27	0.100	0.400	0.350	0.450
Ni/PAN/GCE	13	35	0.100	0.400	0.350	0.450
Ni/PAN-MWCNT/GCE	44	61	0.070	0.385	0.350	0.420

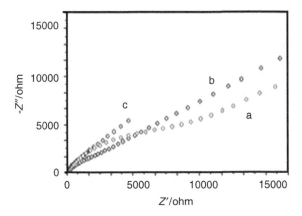

Figure 6.3 Impedance plots of bare GCE (a), PAN/GCE (b) and PAN-MWCNT/GCE (c) in the presence of 1.0 mM $[Fe(CN)_6]^{3-/4-}$ containing 0.1 M KCl as supporting electrolyte.

that the PAN film is a porous material with a large surface area. When MWCNT was immobilized onto PAN, it can be seen that R_{ct} further decreases to about 7.39×10^2 Ω (curve c), implying that the MWCNT can play a very important role similar to an electron conducting tunnel, which makes it easier for the electron transfer to take place. These data showed that the MWCNT and PAN film have been successfully attached to the electrode surface and formed a tunable kinetic barrier.

6.3.3 The surface morphologic studies

The SEM micrographs of PAN/GCE, PAN-MWCNT/GCE and Ni/PAN-MWCNT/GCE are shown in Figure 6.4. The PAN film shows the porous structure with pore diameter of about 200 nm (Figure 6.4a). In general, the size of Ni(II) is in the range of 0.06–0.09 nm. So, Ni (II) ion can be absorbed into the PAN micro pores, which was testified by the experimental result. From Figure 6.4b, it can be seen that the hybrid films are compact, homogeneous and densely packed on the electrode surface and MWCNTs are dispersed on the surface of PAN-MWCNT/GCE. After adsorption of Ni (II) by the immersion method, the micro pores of the PAN film diminished markedly and the film surface was covered partly by Ni (Figure 6.4c). Figure 6.5 gives the EDAX distribution of elements for the deposited Ni. The existence of Ni in the prepared film was confirmed by the EDAX results.

6.3.4 Electrocatalytic behavior of Ni/PAN-MWCNT/GCE nanocomposite

Previous research showed that NiO modified electrodes could display an electrochemical response to glucose (Shamsipur *et al.*, 2010). Here, the electro-catalytic activity of Ni immobilized on Ni/PAN-MWCNT/GCE to carbohydrates was also observed.

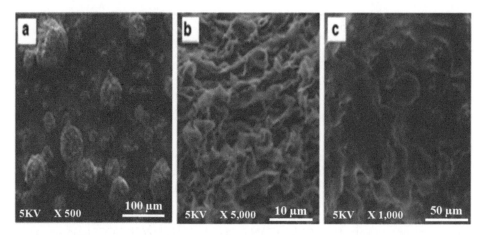

Figure 6.4 SEM images of PAN/GCE (a), PAN-MWCNT/GCE (b) and Ni/PAN-MWCNT/GCE (c) electrodes.

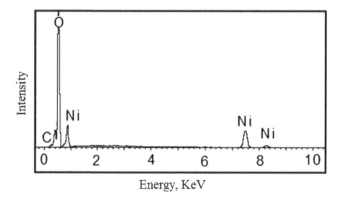

Figure 6.5 EDAX of Ni/PAN-MWCNT/GCE electrode.

As can be seen in Figure 6.6, cyclic voltammograms obtained from bare GCE in 0.1 M NaOH in the absence (curve a) and presence of 1.0 mM carbohydrates (curve b), the oxidation of glucose, sucrose, fructose and sorbitol requires very high positive potentials leading to poorly defined anodic wave due to very slow electrode kinetics. In contrast, oxidation of these carbohydrates at the Ni/PAN-MWCNT/GCE in 0.1 M NaOH occurred at much less positive potential with an increase in the oxidation peak current and a decrease of the reduction peak (curve d). The electrooxidation of carbohydrates due to the existence of Ni (II) ions occurs according to the following reactions:

$$2Ni\ (II) \rightarrow 2Ni\ (III) + 2e^- \tag{6.2}$$

$$2Ni\ (III) + carbohydrate \rightarrow 2Ni\ (II) + radical\ carbohydrate \tag{6.3}$$

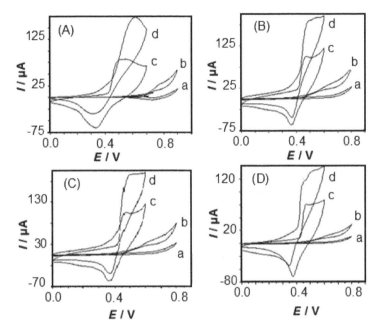

Figure 6.6 Cyclic voltammograms obtained for bare GCE in (a) absence and (b) presence of 1.0 mM of glucose (A), sucrose (B), fructose (C) and sorbitol (D). (c) as (b) and (d) as (b) at Ni/PAN-MWCNT/GCE in 0.1 M NaOH at scan rate 50 mV s^{-1}.

Ni (III) acts as a strong oxidant reacting with the carbohydrate molecules by subtracting a hydrogen atom to yield a radical. Further the reaction of radical with additional surface sites results in product formation. As reported in the literature, gluconolactone and gluconic acid (Tominaga *et al.*, 2005; C. Zhao *et al.*, 2007) were detected as the main products of glucose oxidation. In addition, formate and oxalate (Mho & Johnson, 2001) were reported as oxidation products. Sorbitol was transformed into gluconic acid via the formation of glucose. Glucuronic acid and some degradation side products were also detected (Proença *et al.*, 1997). Sucrose was transformed into α-glucose that changed into β-glucose and converted into gluconic acid (Gülce *et al.*, 1995).

6.3.5 Voltammetry response and calibration curve

The cyclic voltammetric response of carbohydrate electrochemical sensor is investigated by continuous addition various concentrations of carbohydrate into 0.1 M NaOH solution. Figure 6.7 illustrates the cyclic voltammograms and the calibration plot of the response current to different concentrations of glucose. A similar behavior was observed for electrooxidation of sucrose, fructose and sorbitol at Ni/PAN-MWCNT/GCE. The Linear Dynamic Range (LDR), Limit of Detection (LOD) and sensitivity of mentioned carbohydrates was obtained from calibration plots shown in Table 6.2.

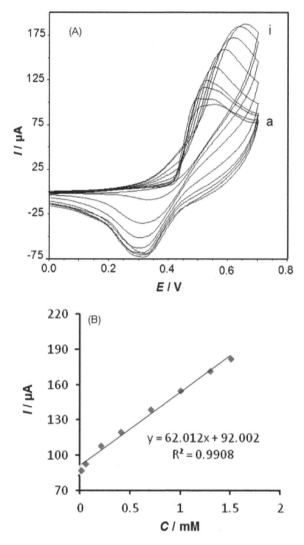

Figure 6.7 (A) Cyclic voltammograms of Ni/PAN-MWCNT/GCE in the presence of a) 0.00, b) 0.01, c) 0.05, d) 0.20, e) 0.40, f) 0.70, g) 1.00, h) 1.30 and i) 1.50 mM of glucose in 0.1 M NaOH solution at scan rate of 50 mV s^{-1}. (B) Plot of anodic peak current vs. glucose concentration.

6.3.6 Interference study

Possible interference in the detection of carbohydrates at the Ni/PAN-MWCNT/GCE was investigated by the addition various compounds such as glutathione, cysteine, tryptophan, glycine, alanine, ascorbic acid, dopamine, uric acid, penicillamine, methionine, lysine, N-acetyl-L-cysteine and cysteamine into 0.1 M NaOH solution in the presence of 1.0 mM carbohydrates. The results showed that the mentioned compounds did not show interference.

Table 6.2 Analytical parameters for voltammetric determination of carbohydrates at different modified electrodes.

Modifier	Sensitivity ($\mu AmM^{-1}cm^{-2}$)	LDR (μM)	LOD (μM)	References
Pt–Pb nanowire	11.25	0–11000	8 (glucose)	(Bai *et al.*, 2008)
OMC	10.81	500–2500	20 (glucose)	(Ndamanisha & Guo, 2009)
Ni(II) oxide/MWCNT	–	200–12000	160 (glucose)	(Shan *et al.*, 2006)
CNT–NiCo-oxide	66.15	10–12120	5 (glucose)	(Arvinte *et al.*, 2011)
CNT–NiCo-oxide	35.72	20–675	9.5 (fructose)	(Arvinte *et al.*, 2011)
Glucose Oxidase	–	4000–15000	– (sucrose)	(Bisenberger, Bräuchle, & Hampp, 1995)
SDH-NADC[a]	–	20–800	40 (sorbitol)	(Saidman, Lobo-Castañón, Miranda-Ordieres, & Tuñón-Blanco, 2000)
Ni/PAN-MWCNT	493.73	10–1500	6 (glucose)	this work
Ni/PAN-MWCNT	384.53	12–3200	7 (sucrose)	this work
Ni/PAN-MWCNT	489.29	7–3500	5 (fructose)	this work
Ni/PAN-MWCNT	358.51	16–4200	11 (sorbitol)	this work

[a] Sorbitol dehydrogenase and nicotinamide adenine dinucleotide

6.3.7 Reproducibility and stability of Ni/PAN-MWCNT/GCE electrode

The reproducibility and storage stability of modified electrode were also examined. The Relative Standard Deviation (RSD) of the sensor response to 1.0 mM carbohydrates was 2.8–5.8% for 5 successive measurements. The RSD for detection of 1.0 mM carbohydrates with four sensors prepared under the same conditions was 3.5–6.2%. When the sensor was stored dry and measured at intervals of 1 week, it retained about 86% of its original sensitivity after 5 weeks.

6.3.8 Determination of glucose and analytical recoveries of carbohydrates

The response of modified sensor to the glucose in human blood serum was investigated. The serum sample obtained from hospitalized patient was analyzed. The results were matched with referenced value obtained from automated standard colorimetric technique in the hospital. Table 6.3 shows analytical recoveries of the glucose, sucrose, fructose and sorbitol solutions added to 0.1 M NaOH solution, and suggesting better accuracy of the method.

6.4 CHAPTER SUMMARY

An electrochemical sensor for carbohydrates detection was fabricated by immobilization of Ni(II) onto the PAN-MWCNT composite via an immersion method. The sensor shows a wide linear range of 10–1500 µM for glucose, 12–3200 µM for sucrose,

Table 6.3 Assay of glucose in human blood serum samples and recovery of carbohydrates in 0.1 M NaOH solution spiked with different concentrations.

Sample	Spiked (mM)	Biochemical analyzer in a local hospital (mM)	Found (mM)	Mean recovery (n = 3)
Blood serum	–	4.70	4.58	–
Glucose	0.15	–	0.14	93 ± 2
Fructose	0.15	–	0.16	107 ± 3
Sorbitol	0.22	–	0.21	95 ± 4
Sucrose	0.31	–	0.30	96 ± 4

7–3500 μM for fructose and 16–4200 μM for sorbitol. In addition, its experimental limit of detection can be achieved as low as 6 μM for glucose, 7 μM for sucrose, 5 μM for fructose and 11 μM for sorbitol. It also possesses good reproducibility and stability. All these advantages can make the designed sensor applicable in medical, food or other areas. Moreover, the investigation also exhibits that the PAN-MWCNT may be applied as a novel immobilization material for a variety of sensor designs.

Chapter 7

Simultaneous detection of D-Penicillamine and L-Tryptophan using ferri/ferrocyanide doped chitosan-multi walled carbon nanotube nanocomposite

7.1 INTRODUCTION

Carbon nanotubes (CNTs) represent an important group of nanomaterials with attractive geometrical, electronic and chemical properties (Kim & Lieber, 1999; Spinks et al., 2002). Since the discovery of CNTs in 1991 (Goyal et al., 2011), considerable efforts have been made to study the application of this new material (Gong et al., 2004; Yuehe Lin et al., 2003; Liu & Lin, 2006; Musameh et al., 2002; Wang et al., 2002; Zhao et al., 2005) but its insolubility in most solvents restrained its application in electroanalysis (Carreño-Gómez & Duncan, 1997). Recently, chitosan (CS) has been reported to disperse functionalized CNTs in aqueous solution with higher efficiency than the organic solvents (Tkac et al., 2007; Zhang et al., 2004).

Chitosan, a natural biopolymer with abundant hydroxyl and amino groups with pK_a 6.5 is soluble in slightly acidic solutions due to the protonation but insoluble in solution above pH 6 due to the deprotonation. Furthermore, the material possesses many special properties such as good adhesion and film forming capability and bio-compatibility. It has been widely used as an immobilization matrix for biosensors and biocatalysis (Qian & Yang, 2006; Maogen Zhang et al., 2006). Negatively charged CNTs and chitosan forms stable and strong interactions due to the polycationic nature of the polymer itself (Yuyang Liu et al., 2005; Wooten & Gorski, 2010; Zhang et al., 2004). Therefore, the CNT–CS composite exhibits robust film-forming ability. In 2009, Pauliukaite et al. studied the immobilization of carbon nanotubes into chitosan film using different cross-linking agents on a graphite–epoxy composite electrode (Pauliukaite et al., 2009).

Penicillamine (2-amino-3-methyl-3-sulfanyl-butanoic acid), PA is a thiol amino acid that exists in D and L enantiomeric forms with different biological and toxicological properties (Wang et al., 2001). This compound is a strong chelating agent and can react with most heavy metal ions. This property is employed for the drug treatment of the Wilson disease, heavy metal intoxication, rheumatoid arthritis and cystinuria (Mazloum-Ardakani et al., 2010). L-Tryptophan (2-amino-3-(1H-indol-3-yl)-propionic acid), L-Trp is the essential amino acid for humans and a precursor for serotonin, melatonin and niacin.

This compound is sometimes added to dietary, food products, pharmaceutical formulas due to its scarcity in vegetables (Walter Kochen, 1994).

Various analytical methods have been reported for the determination of D-PA and L-Trp in both pharmaceutical preparations and biological samples. These methods include high performance liquid chromatography (HPLC) (D'Eramo et al., 1999; D'Eramo et al., 2003; Saetre & Rabenstein, 1978), fluorimetry (Cavrini et al., 1988; Segarra Guerrero et al., 1991), colorimetry (Besada, 1987), spectrophotometry (Al-Majed, 1999; Suliman et al., 2003), chemiluminescence (Viñas et al., 1993), electrochemical methods (Mazloum-Ardakani et al., 2010) and capillary electrophoresis (Wroński, 1996). In general, one of the main problems of these methodologies is the need for previous derivatization of the amino acids but methods based on electroanalysis are the most notable because of their sensitivity, accuracy and simplicity. With respect to relatively large oxidation overpotential of D-PA and L-Trp (Shahrokhian & Bozorgzadeh, 2006; Torriero et al., 2007), the corresponding voltammetric signals on the surface of bare electrodes are usually weak. In order to decrease the undesirable anodic overpotential in the electrochemical oxidation of these compounds and so, propose a specific, direct and selective electrochemical detection, various chemically modified electrodes have been developed in both pharmaceutical formulations and biological samples (Heli et al., 2010; Ojani et al., 2009, 2011; Pournaghi-Azar & Ojani, 1999, 2000; Raoof et al., 2009; Shahrokhian & Fotouhi, 2007; Toito Suarez et al., 2006).

In this research, a chitosan- multi walled carbon nanotube modifier was obtained by using purified MWCNT with chitosan solution (0.5 wt. %), achieving a CS-MWCNT modified glassy carbon electrode (CS-MWCNT/GCE). The chitosan films are permeable to anionic $[Fe(CN)_6]^{3-/4-}$ redox couple, achieving a FC/CS-MWCNT/GC. The FC/CS-MWCNT/GC electrode was used for the simultaneous detection of D-PA and L-Trp, excellent properties of this FC/CS-MWCNT/GCE electrode towards the electrochemical oxidation of D-PA and L-Trp were observed.

7.2 EXPERIMENTAL PROCEDURE

7.2.1 Preparation of FC/CS-MWCNT/GC electrode

Chemical Modified electrode of Ni/MWCNT/GCE, CS-MWCNT/GCE fabricated according to the structure which mentioned in 3.5.4. CS-MWCNT/GCE electrode was placed in an aqueous solution containing 1 mM $[Fe(CN)_6]^{3-/4-}$ and 0.1 M KCl. The electrochemical behavior of the $[Fe(CN)_6]^{3-/4-}$ redox couple was probed by cycling the potential of the CS-MWCNT/GCE electrode between 0.00 V and 0.45 V at a scan rate of 0.05 V/s for 40 complete cycles. CS-MWCNT/GCE electrode was removed from the $[Fe(CN)_6]^{3-/4-}$ solution, rinsed and were cycled in the same potential window in supporting electrolyte (0.1 M KCl) to probe for residual $[Fe(CN)_6]^{3-/4-}$ at the FC/CS-MWCNT/GCE electrode. Then for comparison, FC/GCE, FC/CS/GCE and FC/MWCNT/GCE electrodes were prepared with the same procedure as described above.

7.3   RESULTS AND DISCUSSION

7.3.1 $[Fe(CN)_6]^{3-/4-}$ redox probe

The electrochemical activity of $[Fe(CN)_6]^{3-/4-}$ redox couple at the CS-MWCNT nanocomposite coated gassy carbon electrode was probed using cyclic voltammetry between 0.00 V and 0.45 V (some scans shown in Figure 7.1(A)). As shown in Figure 7.1(A), both cathodic and anodic peak currents gradually increased with subsequent potential scans reaching limiting values after 40 cycles. After 40 potential cycles, the CS-MWCNT coated electrode was removed from the $[Fe(CN)_6]^{3-/4-}$ solution, rinsed briefly with water and placed in 0.1 M KCl. Scanning the same potential window as before and the residual redox activity from $[Fe(CN)_6]^{3-/4-}$ ions trapped within the CS-MWCNT film on the glassy carbon electrode was observed (Figure 7.1(B)). The cathodic and anodic peak current in the first cycle are a factor of three less than the limiting values obtained with $[Fe(CN)_6]^{3-/4-}$ present in the solution.

With increasing number of scans, cathodic and anodic peak currents gradually decay and largely level off after 20 cycles, consistent with the diffusion of $[Fe(CN)_6]^{3-/4-}$ ions out of the film. The persistent $[Fe(CN)_6]^{3-/4-}$ signal after repeated cycling in electrolyte solution indicates that residual $[Fe(CN)_6]^{3-/4-}$ ions are trapped in the CS-MWCNT film. Average $E_{1/2}$ and ΔE values obtained for the residual $[Fe(CN)_6]^{3-/4-}$ redox species at the CS-MWCNT coated electrode were 243 mV and 65 mV, respectively.

Diffusion coefficient values of $[Fe(CN)_6]^{3-/4-}$ are calculated from the scan rate dependence of the peak current height of CVs recorded at GCE, CS/GCE, MWCNT/GCE and CS-MWCNT/GCE electrodes over a scan rate range of 10–300 mV/s. Apparent diffusion coefficients (D_{app}) are determined from the slopes of the plots of peak current versus $v^{1/2}$. Diffusion coefficients calculated from this treatment were 1.9×10^{-6}, 3.8×10^{-6}, 1.7×10^{-6} and 5.6×10^{-6} cm^2/s at glassy carbon electrode, CS/GCE, MWCNT/GCE and CS-MWCNT/GCE electrode, respectively. These apparent diffusion coefficient values for $[Fe(CN)_6]^{3-/4-}$ are comparable to those reported by another researchers for chitosan films deposited at gold electrodes (Zangmeister et al., 2006).

7.3.2   Characterization of CS-MWCNT/GCE electrode

CV and EIS were used to characterize the modification of the electrode in 2 mM $[Fe(CN)_6]^{3-/4-}$ and 0.1 M KCl solution. Figure 7.2 compares the CV responses at GCE, CS/GCE, MWCNT/GCE and CS-MWCNT/GCE electrodes in the above solution, respectively. The defined peak currents occurred at the CS/GCE electrode (curve b) is due to the good permeability and preconcentration of $[Fe(CN)_6]^{3-/4-}$ in the CS film.

After modification of GC electrode with MWCNT film, the anodic peak and cathodic peak current is increased indicating the MWCNTs can improve the electroactive surface area of the electrode (curve c). When CS-MWCNT composite was modified on the glassy carbon electrode surface, the peak current increased distinctly due to the formation of conduction pathway in the CS-MWCNT composite film (curve d).

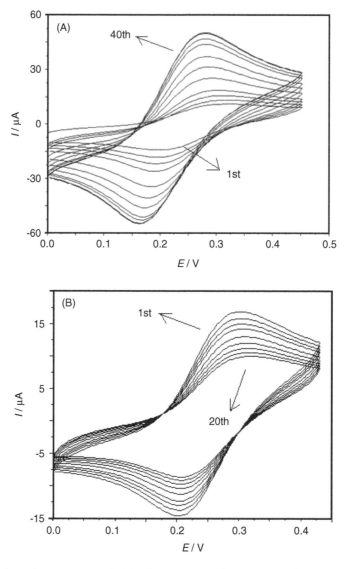

Figure 7.1 Cyclic voltammograms of (A) 40 cycles of CS-MWCNT/GCE electrode in 1.0 mM [Fe(CN)$_6$]$^{3-/4-}$ and 0.1 M KCl at 0.05 V/s and (B) 20 cycles of CS-MWCNT/GCE electrode post [Fe(CN)$_6$]$^{3-/4-}$ exposure in 0.1 M KCl.

After modification of glassy carbon electrode with MWCNT film, the anodic peak and cathodic peak current is increased indicating the MWCNTs can improve the electroactive surface area of the electrode (curve c). When CS-MWCNT composite was modified on the glassy carbon electrode surface, the peak current increased distinctly due to the formation of conduction pathway in the CS-MWCNT composite film (curve d).

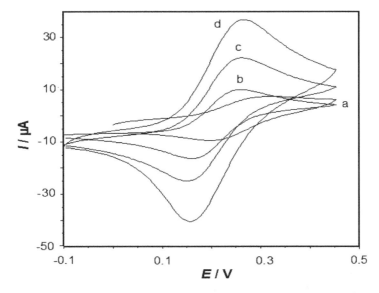

Figure 7.2 The cyclic voltammogram of (a) GCE, (b) CS/GCE, (c) MWCNT/GCE and (d) CS-MWCNT/GCE electrodes in 2.0 mM $[Fe(CN)_6]^{3-/4-}$ and 0.1 M KCl as supporting electrolyte at a scan rate 0.05 V/s.

Electrochemical impedance spectroscopy was also used to identify the different modifications on the glassy carbon electrode. Figure 7.3(a–c) shows the impedance spectra of GCE, CS/GCE and CS-MWCNT/GCE electrodes, respectively. As shown in Figure 7.3(a), unmodified glassy carbon electrode, the diameter of the semicircle; the charge-transfer resistance (R_{ct}) was estimated to be $1.1 \times 10^4 \, \Omega$. As for CS/GCE, the R_{ct} value decreased to $4.2 \times 10^3 \, \Omega$ due to positively charged chitosan promoting electron transfer of negatively charged $[Fe(CN)_6]^{3-/4-}$ couple to the electrode surface (Figure 7.3(b)). After modification of GCE with CS-MWCNT, the R_{ct} value decreased to $7.1 \times 10^2 \, \Omega$ indicating MWCNT could form high electron transfer bridges between the electrode and electrolyte (Figure 7.3(c)). These results presented in Figure 7.3 were found in a good agreement with the results obtained by using CV.

The surface morphologies of unmodified and modified electrodes were examined by using SEM (Figure 7.4(a–d)). As shown in Figure 7.4a, after modification of glassy carbon with chitosan, some parts of glassy carbon became invisible due to chitosan immobilization onto the electrode surface. In the SEM image of CS-MWCNT modified glassy carbon electrode, the substrate was mostly covered with homogeneous MWCNT films (Figure 7.4(c)). From Figure 7.4(d), it can be seen that the hybrid films are compact, homogeneous and densely packed on the electrode surface and FC particles are dispersed on the surface of CS-MWCNT/GCE. Whereas, the chitosan wrapped on MWCNT is positively charged because of $-NH_3^+$ groups existed within the chitosan; therefore, resulting in a strong electrostatic interaction between the FC and CS-MWCNT film.

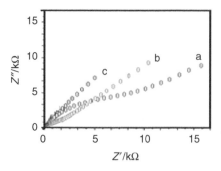

Figure 7.3 Nyquist plots recorded at (a) bare GCE, (b) CS/GCE and (c) CS-MWCNT/GCE electrode in the presence of 1.0 mM $[Fe(CN)_6]^{3-/4-}$ solution containing 0.1 M KCl as supporting electrolyte.

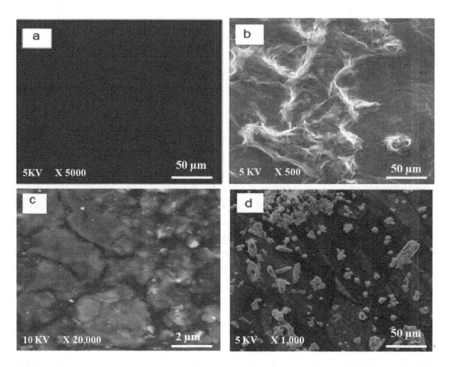

Figure 7.4 SEM images of (a) GCE, (b) CS/GCE, (c) CS-MWCNT/GCE and (d) FC/CS-MWCNT/GCE electrodes.

As shown in UV-Vis absorption spectrum (Figure 7.5), CS had an absorption peak at 230 nm in 1% acetic acid (curve a). After mixing the CS with MWCNT, the absorption value increased whereas the peak at 230 nm remains the same (curve b). This observation indicates that, the chitosan could be readily coupled with the side wall of acid-treated carbon nanotubes through the interactions among the carboxyl groups of the nanotubes and the NH_3 group of chitosan.

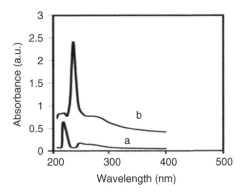

Figure 7.5 UV-Vis spectra of (a) chitosan in 0.1 M acetic acid solution mixture and (b) mixture acidic solution of chitosan and multi walled carbon nanotube.

Table 7.1 The Anodic and cathodic potential, the half-wave potential and the peak-to-peak separation potential for redox couple of cyclic voltammogram of Figure 7.2.

Electrode	I_{pc} (μA)	I_{pa} (μA)	ΔE_p (V)	$E_{1/2}$ (V)	E_{pc} (V)	E_{pa} (V)
GCE	−10	3	0.100	0.250	0.200	0.300
CS/GCE	−18	10	0.090	0.205	0.160	0.250
MWCNT/GCE	−25	20	0.100	0.200	0.150	0.250
CS-MWCNT/GCE	−40	35	0.100	0.200	0.150	0.250

Anodic and cathodic potential, the half-wave potential and also the peak-to-peak separation potential for redox couple in each of the modified electrode was measured according to the reference electrode potential in cyclic voltammogram of Figure 7.2 and presented in Table 7.1. It is clear that CS-MWCNT/GCE reduced the overpotential.

7.3.3   Characterization of FC/CS-MWCNT/GC electrode in the presence of *D*-PA and *L*-Trp

The electrocatalytic oxidation of 0.1 mM *D*-PA and 0.1 mM *L*-Trp was investigated in a 0.1 M phosphate buffer solution (pH 7) at the bare GCE (Figures 7.6A and 7.6B, curve b) and FC/CS-MWCNT/GCE (Figure 7.6A, curve d and Figure 7.6B, curve e) electrodes. As shown in the CVs (Figure 7.6), the electrocatalytic oxidation of *D*-PA can be catalyzed by the ferric (Fe^{3+}) preconcentrated in CS-MWCNT/GCE film or produced from electrooxidation of preconcentrated ferrous (Fe^{2+}) in film as mediator and the oxidation peak potential of *D*-PA shifted almost 500 mV to the less positive potential at FC/CS-MWCNT/GCE electrode. The following mechanism (EC) (chemical reactions 7.1–7.3) is represented for the electrocatalytic oxidation of *D*-PA (RSH). However, FC cannot catalyze electrooxidation of *L*-Trp and the oxidation peak

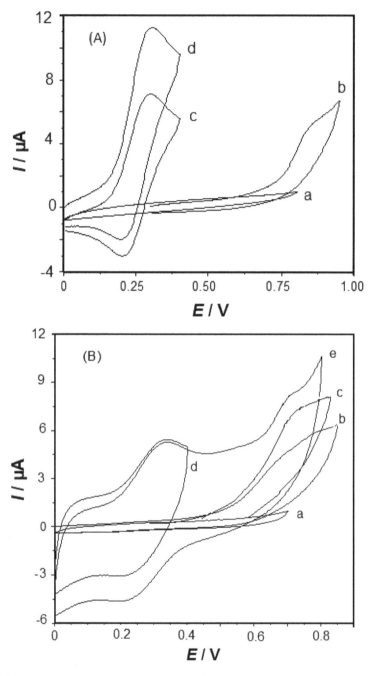

Figure 7.6 Cyclic voltammograms of (A) pH 7.00 PBS in (a) absence and (b) presence of 0.1 mM of D-PA at GCE and (c) as (a) and (d) as (b) at FC/CS-MWCNT/GCE electrode at scan rate of 0.02 V/s; (B) pH 7.00 PBS in (a) absence and (b) presence of 0.1 mM of L-Trp at GCE, (c) in presence of 0.3 mM D-PA + 0.3 mM L-Trp mixture at CS-MWNT/GCE, (d) as (a) and (e) as (b) at FC/CS-MWCNT/GC electrode.

potential of L-Trp did not shift. The result shows that, at the bare glassy carbon electrode, oxidation of D-PA and L-Trp occurs with a peak potential of nearly 800 mV vs. Ag|AgCl|KCl$_{3M}$.

$$RSH + H_2O \rightarrow RS^- + H_3O^+ \tag{7.1}$$

$$2Fe^{2+} \rightarrow 2Fe^{3+} + 2e^-(E) \tag{7.2}$$

$$2RS^- + 2Fe^{3+} \rightarrow RS - SR + 2Fe^{2+}(C) \tag{7.3}$$

7.3.4   Effect of pH on *D*-PA and *L*-Trp oxidation

It is well known that, the electrochemical behavior of D-PA and L-Trp are dependent on the pH value of the aqueous solution (Shahrokhian & Fotouhi, 2007; Toito Suarez *et al.*, 2006). The effect of pH value on the electrooxidation of mentioned compounds on the surface of FC/CS-MWCNT/GCE electrode was investigated through the use of different 0.1 M phosphate buffer solutions (pH 3–9). The current response of D-PA and L-Trp at FC/CS-MWCNT/GCE electrode increases from pH 3 to 7, and then decrease at pH values higher than 7 (not shown). Therefore, pH 7 was chosen as optimum pH and further studies were performed at pH 7.

7.3.5   Simultaneous detection of *D*-PA and *L*-Trp

The next attempt was taken to detect D-PA and L-Trp simultaneously by using the FC/CS-MWCNT/GCE electrode. Figure 7.7A shows the DPVs (Differential Pulse Voltammetry) obtained at the modified electrode when analytical experiments were carried out by varying the concentration of D-PA in the presence of fixed amount of L-Trp (100 μM) and varying L-Trp concentration in the presence of fixed amount of D-PA (3 μM) at pH 7and scan rate of 20 mVs^{-1}. The presence of FC film on CS-MWCNT composite resolved the mixed voltammetric response into two well-defined voltammetric peaks at potentials 0.25 and 0.70 V, corresponding to the oxidation of D-PA and L-Trp, respectively. The separation between the two peak potentials is sufficient enough for the simultaneous determination of D-PA and L-Trp. DPVs results indicate that the calibration curves for D-PA and L-Trp were linear for the whole concentrations range investigated (3–300 μM for D-PA and 7–300 μM for L-Trp) with good correlation coefficients. Experimental results showed that the detection limit was 0.9 μM for D-PA and 4.0 μM for L-Trp (S/N = 3). The obtained analytical parameters for D-PA and L-Trp determination at the surface of modified electrode are comparable with the results reported by other research groups is represented in Table 7.2.

7.3.6   Stability and reproducibility of the FC/CS-MWCNT/GC electrode

After storage period of two weeks in pH 7 the PBS slightly changed the currents for the responses to D-PA and L-Trp. For example, the FC at FC/CS-MWCNT/GCE electrode can maintain 90% of its initial response after two weeks. Thus, CS-MWCNT composite is very efficient to retain the activity of adsorbed FC. The fabrication and

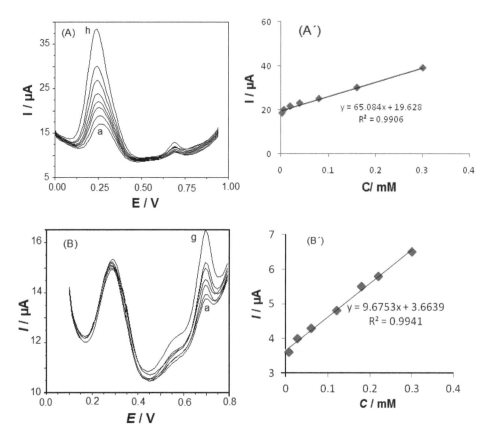

Figure 7.7 DPVs of (A) 100 µM Trp in presence of different concentrations of D-PA: (a) 0.000, (b) 0.003, (c) 0.007, (d) 0.020, (e) 0.040, (f) 0.080, (g) 0.160 and (h) 0.300 mM and (B) 3.0 µM D-PA in presence of different concentrations of Trp: (a) 0.007, (b) 0.028, (c) 0.060, (d) 0.120, (e) 0.180, (f) 0.220 and (g) 0.300 mM at the FC/CS-MWCNT/GC electrode in pH 7.00 PBS. (A′) and (B′) are the Plot of anodic peak currents vs. D-PA and Trp concentrations, respectively.

reproducibility of three electrodes made independently showed an acceptable reproducibility with the relative standard deviations of 5.1% and 3.8% for the current determinations of 10 µM D-PA and 50 µM L-Trp, respectively.

7.3.7 Real sample analysis

For the purpose of proving its practical applications, the modified electrode was used to determine the content of D-PA in the tablet using DPVs. A standard addition method was adopted to estimate the accuracy and the measurement results as shown in Table 7.3. The recovered ratios of D-PA and L-Trp were investigated, suggesting better accuracy of the method.

Table 7.2 Analytical parameters for voltammetric determination of *D*-PA at different modified electrode.

Electrode	Modifier	Shifted potential value (mV)	LDR (μM)	LOD (μM)	Reference
CP	Ferrocene carboxylic acid	420	6.500–100	6.150	(Raoof et al., 2007)
GC	Dopamine	–	0.100–2.500	0.080	(Shahrokhian & Bozorgzadeh, 2006)
GC	Tyrosinase	–	0.020–80	0.007	(Torriero et al., 2006)
CP	Quinizarine	220	0.800–140	0.760	(Mazloum-Ardakani et al., 2010)
GC	FC/CS-MWCNT	500	3–300	0.900	This work

Table 7.3 Measurement results of *D*-PA in commercial tablet (Dr. Abidi pharmaceutical laboratories Co.) and recovery of *D*-PA and Trp in 0.1 M phosphate buffer solution spiked with different concentrations.

Sample	Added (mM)	Found (mM)	Mean recovery (n = 3)
Tablet *D*-PA	0.0080	0.0077	–
D-PA	0.0100	0.0095	95 ± 3
	0.0500	0.0510	108 ± 2
Trp	0.1000	0.0960	96 ± 3
	0.0700	0.0600	94 ± 4

7.4 CHAPTER SUMMARY

CS-MWCNT film has been successfully deposited on glassy carbon electrode. Subsequently, FC can be adsorbed onto this film and realize its electron transfer. The uniform CS-MWCNT film showed great enhancement of FC loading and improvement of its behavior. The fabricated sensor displays good stability and acceptable reproducibility. The FC/CS-MWCNT nanocomposite also provides a promising platform for the simultaneous determination of *D*-PA an *L*-Trp with high sensitivity and a low detection limit. Furthermore, FC/CS-MWCNT presented excellent selectivity and satisfying recoveries in detecting *D*-PA from *D*-PA tablets. In conclusion, the FC/CS-MWCNT/GCE electrode is simple and practical. It is a good electrochemical sensor for direct determination of *D*-PA and *L*-Trp.

Chapter 8

Outlook of nanocomposites in electrochemical sensors

Carbon Nanotubes (CNTs) have become the subject of intense researches in the last decades because of their unique properties and the promising applications in any aspect of nanotechnology. Because of their unique one-dimensional nanostructures, CNTs display fascinating electronic and optical properties that are distinct from other carbonaceous materials and nanoparticles of other types. CNTs are widely used in electronic and optoelectronic, biomedical, pharmaceutical, energy, catalytic, analytical, and material fields. Particularly, the properties of small dimensions, functional surfaces, good conductivity, excellent biocompatibility, modifiable sidewall, and high reactivity make CNTs ideal candidates for constructing sensors with high performances. As an example, CNTs have been extensively employed in constructing various electrochemical sensors. Compared with the conventional scale materials and other types of nanomaterials, the special nano-structural properties make CNTs have some overwhelming advantages in fabricating electrochemical sensors, including,

(i)   the large specific area producing high sensitivity;
(ii) the tubular nanostructure and the chemical stability allowing the fabrication of ultrasensitive sensors consisting of only one nanotube;
(iii) the good biocompatibility that is suitable for constructing electrochemical biosensors, especially for facilitating the electron transfer of redox proteins and enzymes;
(iv) the modifiable ends and sidewalls providing a chance for fabricating multifunctioned electrochemical sensors via the construction of functional nanostructures;
(v)   the possibility of achieving miniaturization;
(vi) the possibility of constructing ultrasensitive nanoarrays.

Generally, the replacement of ordinary materials by CNTs can effectively improve the redox currents of inorganic molecules, organic compounds, macrobiomolecules or even biological cells and reduce the redox overpotentials (Bagheri *et al.*, 2014b). The electron transfer and the direct electrochemistry of redox proteins at CNT-based electrochemical sensors were also widely reported. Due to the well-defined structure, the chemistry stability and the electrocatalytic activity toward many substances, CNTs are also extensively used as the carrier platforms for constructing various electrochemical sensors (Chekin *et al.*, 2014b).

Many of the carbon nanotube based technologies are promising for use as biosensors. These technologies are expanding rapidly, as more and more researchers are discovering the benefits of incorporating CNTs into sensors. However, CNT biosensors have yet to achieve the performance that has been speculated based on the catalytic and electronic properties of the material (Arab *et al.*, 2014). There are still numerous challenges to implementing CNT into active biosensors that can be widely used. These include: a cost-effective method for separating different types of nanotubes, the difficulty in miniaturization of sensors, possible toxicity of nanotubes, and the technological difficulty in fabrication (Julkapli *et al.*, 2015b).

In this book, metal oxide based nanoparticles, i.e. zinc oxide and titanium dioxide (ZnO, TiO$_2$) by a simple and low-cost sol-gel method using gelatin as an organic precursor was synthesized. The synthesized nanomaterials are characterized by thermogravimetric analysis (TGA), powder X-ray diffraction (XRD), transmission electron microscopy (TEM), Fourier transform infrared spectroscopy (FT-IR) and ultraviolet visible (UV-Vis) spectroscopy techniques for their structure assessment. The obtained results indicate that ZnO nanoparticles have high crystallinity with hexagonal structure. Gelatin plays a very important role in the formation of zinc oxide and titanium dioxide nanoparticles. Long-chain gelatin compounds were used to terminate the growth of nanoparticles and to stabilize them, it expands during the calcination process so the particles cannot come together easily that's why it can prevent agglomeration. The nanocomposite film is prepared using the as-synthesized ZnO nanoparticles together with multi walled carbon nanotube and polycaprolactone on glassy carbon electrode (ZnO/MWCNT/PCL/GCE).

The porous ZnO/MWCNT/PCL film was used as a supporting matrix to immobilize Co(II) ions. Immobilized Co(II) ions exhibit excellent electrocatalytic activities to glucose electrocatalysis oxidation. The sensor responded linearly to glucose in the concentration of 5.0×10^{-5} to 6.0×10^{-3} M with detection limit of 1.6×10^{-5} M at 3σ using cyclic voltammetric measurements. The sensor could retain the electrochemistry of immobilized Co(II) at constant current values upon the continuous CV sweep over the potential range from 0.15 V to 0.4 V at 50 mV s^{-1}, suggesting that the Co(II) can tightly adsorb on the surface of ZnO/MWCNT/PCL-modified GC electrode. When stored for over 2 weeks, the sensor retained 90% of the initial sensitivity to glucose. The relative standard deviation of the sensor was a 4.7% for 15 successive determinations in 1.0 mM glucose solution, indicating that it is a good precision so it exhibited good reproducibility and long-term stability. Also, the fabricated electrode displays a voltammetric response to the glucose in human blood serum and the results were matched with referenced value obtained by the standard technique. In addition, the sensor based on the Co/ZnO/MWCNT/PCL composite film modified electrodes exhibited good electrocatalytic response to the oxidation of glucose, good reproducibility and stability. Therefore, this ZnO/MWCNT/PCL composite film could offer a new promising platform for further study of the direct electrochemistry of redox proteins and the development of biosensors.

On the other hand, the anatase titanium dioxide nanoparticles are successfully synthesized by sol-gel method with and without gelatin. The titanium dioxide nanoparticles were structurally characterized by analytical and spectroscopic techniques, such as thermogravimetric analysis (TGA), X-ray diffraction (XRD), transmission electron microscopy (TEM), fourier transform infrared spectroscopy (FTIR) and ultraviolet visible spectroscopy (UV-Vis) techniques. The particle sizes of synthesized titanium dioxide nanoparticles with and without gelatin were $\sim$13 nm and $\sim$21 nm, respectively.

The main advantage of using gelatin is that, it acts as a stabilizing agent and it provides a long-term stability and also it can prevent the agglomeration of nanoparticles during preparation. The results indicate that the gelatin is a reliable green stabilizer, which can be used as a polymerization agent in the sol-gel method.

Additionally, the composite thin film was prepared by synthesized and multi walled carbon nanotube on glassy carbon electrode (TiO_2-MWCNT/GCE). The TiO_2-MWCNT/GCE responded linearly to L-Tryptophan in the concentration of 1.0×10^{-6} to 1.5×10^{-4} M with a detection limit of 5.2×10^{-7} M at 3σ using amperometry. The electrochemical sensor studies exhibits the good reproducibility as well as long-term stability. This demonstration indicates that, this is a new method of synthesizing the gelatin based TiO_2/MWCNT composite sensors, catalytic bioreactors and biomedical devices using gelatin.

At this stage, a Ni(II) incorporation into porous polyacrylonitrile- multi walled carbon nanotubes composite modified glassy carbon electrode (Ni/PAN-MWCNT/GCE) was fabricated by simple drop-casting and immersion technique. The unique electrochemical activity of Ni/PAN-MWCNT composite modified glassy carbon electrode was illustrated in 0.10 M NaOH using cyclic voltammetry. The Ni/PAN-MWCNT/GCE exhibits the characteristic of improved reversibility and enhanced current responses of the Ni(II)/Ni(III) couple compared to Ni/PAN/GCE and Ni/MWCNT/GCE. The results of electrochemical impedance spectroscopy (EIS) and scanning electron microscopy (SEM) indicated the successful immobilization for PAN-MWCNT composite film. Kinetic parameters such as the electron transfer coefficient (α) and rate constant (k_s) of the electrode reaction were determined. The Ni/PAN-MWCNT/GCE also shows very good electrocatalytic activity toward the electrocatalytic oxidation of carbohydrates (glucose, sucrose, fructose and sorbitol).

The electrocatalytic response showed a wide linear range (10–1500 μM, 12–3200 μM, 7–3500 μM and 16–4200 μM for glucose, sucrose, fructose and sorbitol, respectively) as well as its experimental limit of detection can be achieved 6 μM, 7 μM, 5 μM and 11 μM for glucose, sucrose, fructose and sorbitol, respectively. The modified electrode for carbohydrates determination benefits for its simple preparation method, good electrode stability and high electrode sensitivity. The investigation exhibits that the nanocomposite of PAN-MWCNT could be applied as a novel immobilization material for a variety of sensor designs.

In this contribution, we report a simple and effective strategy for the fabrication of nanocomposite containing chitosan and multi walled carbon nanotube coated on a glassy carbon electrode. The characterization of the modified electrode (CS-MWCNT/GCE) was carried out using (SEM) and absorption spectroscopy (UV-Vis). The electrochemical behavior of CS-MWCNT/GCE electrode was investigated and compared with the electrochemical behavior of chitosan modified glassy carbon electrode (CS/GCE), multi walled carbon nanotube modified glassy carbon electrode (MWCNT/GCE) and unmodified glassy carbon electrode using cyclic voltammetry and electrochemical impedance spectroscopy.

The chitosan films are electrochemically inactive; similar background charging currents are observed at bare glassy carbon electrode. The chitosan films are permeable to anionic $[Fe(CN)_6]^{3-/4-}$ redox couple. Electrochemical parameters including apparent diffusion coefficient for the $[Fe(CN)_6]^{3-/4-}$ redox probe at FC/CS-MWCNT/GCE electrode is comparable to values which reported for casted chitosan films. This modified

Table 8.1 Summary of the electrochemical sensors properties which were fabricated in this research.

Catalytic Function	Catalyst promoter/support	Analyte	LDR (M)	LOD (M)
Co (II)	ZnO/MWCNT/ PCL/GCE	Glucose	5.0×10^{-5} to 6.0×10^{-3} M	$1.6{-}10^{-5}$ M
TiO$_2$	MWCNT/GCE	L-Tryptophan	1.0×10^{-6} to 1.5×10^{-4} M	5.2×10^{-7} M
Ni (II)	PAN/MWCNT/ GCE	Carbohydrates (glucose, sucrose, fructose and sorbitol)	10–1,500, 12–3,200, 73,500, and 16–4,200 μM	6, 7, 5, and 11 μM
$[Fe(CN)_6]^{3-/4-}$	CS-MWCNT/ GCE	D-Penicillamine and L-Tryptophan	3–300 μM for D-PA and 7–300 μM for L-Trp	0.9 μM and 4.0 μM

electrode showed electrocatalytic effect for the electrocatalytic oxidation and simultaneous determination of *D*-penicillamine and *L*-Tryptophan. The detection limit of 0.9 μM and 4.0 μM for *D*-penicillamine and *L*-Tryptophan, respectively, makes this nanocomposite very suitable for determination of them with good sensitivity.

One of the main objectives of this research is to develop the electrocatalytic oxidation in electrochemical sensors by using metal oxide nanoparticles, conductive polymers and multi walled carbon nanotubes by chemically modified electrodes. Conducting a series of experiments by using synthesized metal oxide nanomaterials shows that, the nanomaterials are very good candidates for developing the electrochemical sensors. The results of this book are summarized in the table below. In the broader picture the aim of this research work is to fabricate a good catalyst nanocomposite film for chemically modified glassy carbon electrode and make a good electrochemical sensor which is low cost, good efficiency, eco-friendly, sufficient selectivity, high sensitivity, accuracy, high controllable quality, reproducibility and reliability.

More economical ways to grow CNTs as well as to connect them to electrical contacts must be developed (Zamiri *et al.*, 2015). The utilization of electric forces to grow carbon is not cost-effective and cannot be used for mass production. NanoIntegris Inc., USA has developed a proprietary technology to develop electronically pure metallic and semiconducting SWCNTs. However, still more research efforts are required to arrive at highly pure CNTs having greater than 99.99% purity as even the minute amounts of impurities may impact their behavior. Therefore, based on the challenges in the technology, CNTEC sensing technology is not ripe enough to make realistic commercial products as it cannot fulfil the desired quality control standards posed by Food and Drug Administration (FDA) and other certification agencies (Bagheri & Ziglari, 2015).

There have been continuously increasing reports for the development of CNT based EC sensors in the past decade and most of the researchers and companies are progressing towards the protection of their intellectual property in this area as shown by the increasing number of patent applications filed. As the field of CNT based EC sensors is evolving and maturing very rapidly, it is most likely that the existing challenges in the commercialization of these sensors would be effectively tackled by the technology developments in near future (Bagheri *et al.*, 2015d; Kargar *et al.*, 2015).

The continuous development in this area would further lead to cost-effective and reproducible manufacturing techniques; methods for electrode improvement; miniaturized prototypes with lesser power consumption; and, robust, reproducible and highly sensitive sensors (Babadi *et al.*, 2015). Last but not least, the current market price of CNTs is also too high presently for any realistic commercial application. In the last few years, there have been several companies that were set up in many different countries, particularly the USA and China, to produce and market nanotubes. It remains to be seen whether CNTs will be available to consumers for less than US $100/kg. Currently, single walled CNTs with over 90% purity are marketed for almost US $40,000/kg compared to $US 2000–3000/kg for multi-walled CNTs with purity over 95%.

The chapter cannot be concluded without a slight cautionary note. CNTs might not be an answer for all applications in electrochemical sensing and nanobiotechnology (Chekin *et al.*, 2015a). Other carbon-based nanomaterials such as doped diamond like-materials and graphene have emerged and also opened new avenues toward the development of new and/or improved electrochemical electrochemical sensing schemes. Despite biosensors have been developed for numerous analytes, glucose biosensors are still a highly active area of biosensor research and account for over 80% of the electrochemical sensor industry (Amin TermehYousefi, 2015). Besides glucose, recent electrochemical research has focused on the detection of neurotransmitters such as dopamine and its derivatives. Of further interest is the detection of glycated hemoglobin, a biomarker used to follow long-term blood glucose level (Yousefi *et al.*, 2014).

The design and fabrication of electrochemical biosensors using CNTs together with nanoscale materials remains a vibrant research area (Chekin *et al.*, 2015b). Glucose sensing will be still the most important research area considering over 10 billion glucose assays need to be performed by diabetic people. However, for commercial viability CNT-based sensors must be manufactured with high consistency and cost-effectiveness so that they can compete with other commercially available products. It is a formidable task to translate research findings into product lines and doubtlessly some endeavours will succeed while some will fail miserably (Julkapli *et al.*, 2015a). One of the main reasons is the exorbitant cost of introducing a new device to compete with existing technologies (Yousefi *et al.*, 2012). As an example, the cost of introducing a continuous glucose monitoring system is over US100 million, which is beyond the means of a small enterprise.

References

Abbas, Z., Holmberg, J. P., Hellström, A. K., Hagström, M., Bergenholtz, J., Hasellöv, M., & Ahlberg, E. (2011). Synthesis, characterization and particle size distribution of TiO_2 colloidal nanoparticles. *Colloids and Surfaces A: Physicochemical and Engineering Aspects, 384*(1–3), 254–261. doi: 10.1016/j.colsurfa.2011.03.064

Abdolvahabi, Z., Bagheri, S., Haghighi, S., & Karimi, F. (2012a). Relationship between Emotional Intelligence and Self-efficacy in Practical courses among Physical Education Teachers. *European Journal of Experimental Biology, 2*(5), 1778–1784.

Abdolvahabi, Z., Bagheri, S., & Kioumarsi, F. (2012b). Relationship between Emotional Intelligence and Self-efficacy in Research among Tehran Physical Education Teachers. *European Journal of Experimental Biology, 2*(6), 2337–2343.

Advani, S. G. (2007). *Processing and properties of nanocomposites*: World Scientific.

Ajayan, P. M., Schadler, L. S., & Braun, P. V. (2006). *Nanocomposite science and technology*: Wiley-Vch.

Al-Majed, A. A. (1999). Spectrophotometric estimation of d-penicillamine in bulk and dosage forms using 2,6-dichloroquinone-4-chlorimide (DCQ). *Journal of Pharmaceutical and Biomedical Analysis, 21*(4), 827–833. doi: 10.1016/s0731-7085(99)00215-0

Alivisatos, A. P. (1996). Perspectives on the Physical Chemistry of Semiconductor Nanocrystals. *The Journal of Physical Chemistry, 100*(31), 13226–13239. doi: 10.1021/jp9535506

Almquist, C. B., & Biswas, P. (2002). Role of synthesis method and particle size of nanostructured TiO_2 on its photoactivity. *Journal of Catalysis, 212*(2), 145–156.

Amin TermehYousefi, S. B., Nahrizul Adib Kadri ,Mohamad Rusop Mahmood, Shoichiro Ikeda. (2015). Controlling Vaporization Time As Effective Parameter on Purified Vertically Aligned Carbon Nanotubes Based on CVD Method. *Fullerenes, Nanotubes and Carbon Nanostructures.*

Amir, I., Nur, M., Julkapli, N. M., Bagheri, S., & Yousefi, A. T. (2015). TiO_2 hybrid photocatalytic systems: impact of adsorption and photocatalytic performance. *Reviews in Inorganic Chemistry, 35*(3), 151–178.

Amiri, O., Salavati-Niasari, M., Farangi, M., Mazaheri, M., & Bagheri, S. (2015). Stable Plasmonic-Improved dye Sensitized Solar Cells by Silver Nanoparticles Between Titanium Dioxide Layers. *Electrochimica Acta, 152*, 101–107.

Amiri, O., Salavati-Niasari, M., Hosseinpour-Mashkani, S. M., Rafiei, A., & Bagheri, S. (2014). Cadmium selenide@ sulfide nanoparticle composites: Facile precipitation preparation, characterization, and investigation of their photocatalyst activity. *Materials Science in Semiconductor Processing, 27*, 261–266.

Arab, M. M., Yadollahi, A., Hosseini-Mazinani, M., & Bagheri, S. (2014). Effects of antimicrobial activity of silver nanoparticles on in vitro establishment of G× N15 (hybrid of almond× peach) rootstock. *Journal of Genetic Engineering and Biotechnology, 12*(2), 103–110.

Arvinte, A., Sesay, A.-M., & Virtanen, V. (2011). Carbohydrates electrocatalytic oxidation using CNT–NiCo-oxide modified electrodes. *Talanta, 84*(1), 180–186. doi: 10.1016/j.talanta.2010.12.051

Azizian, J., Shameli, A., Balalaie, S., Mehdi Ghanbari, M., Zomorodbakhsh, S., Entezari, M., *et al.* (2012). The one-pot synthesis of pyrano [2, 3-d] pyrimidinone derivatives with 1, 4-diazabicyclo [2.2. 2] octane in aqueous media. *Oriental Journal of Chemistry, 28*(1), 327.

Babadi, A. A., Bagheri, S., & Hamid, S. B. A. (2015). Progress on antimicrobial surgical gloves: a review. *Rubber Chemistry and Technology*.

Bagheri, S., Chandrappa, K., & Hamid, S. B. A. (2013a). Generation of Hematite Nanoparticles via Sol-Gel Method. *Research Journal of Chemical Sciences. ISSN, 2231*, 606X.

Bagheri, S., Chekin, F., & Hamid, S. B. A. (2014a). Cobalt Doped Titanium Dioxide Nanoparticles: Synthesis, Characterization and Electrocatalytic Study. *Journal of the Chinese Chemical Society, 61*(6), 702–706.

Bagheri, S., Chekin, F., & Hamid, S. B. A. (2014b). Gel-assisted synthesis of anatase TiO_2 nanoparticles and application for electrochemical determination of L-tryptophan. *Russian Journal of Electrochemistry, 50*(10), 947–952.

Bagheri, S., Hassani, S., & Naderi, G. (2012a). Theoretical study on physicochemical and geometrical properties of Doxorubicin and Daunorubicin conjugated to PEO-b-PCL nanoparticles. *European Journal of Experimental Biology, 2*(3), 641–645.

Bagheri, S., Hir, Z. A. M., Yousefi, A. T., & Hamid, S. B. A. (2015a). Progress on Mesoporous Titanium Dioxide: Synthesis, Modification and Applications. *Microporous and Mesoporous Materials*.

Bagheri, S., Julkapli, N. M., & Hamid, S. B. A. (2014c). Functionalized activated carbon derived from biomass for photocatalysis applications perspective. *Int. J. Photoenergy*.

Bagheri, S., Julkapli, N. M., & Yehye, W. A. (2015b). Catalytic conversion of biodiesel derived raw glycerol to value added products. *Renewable and Sustainable Energy Reviews, 41*, 113–127.

Bagheri, S., KG, C., & Hamid, S. B. A. (2013b). Facile synthesis of nano-sized ZnO by direct precipitation method. *Der Pharma Chemica, 5*(3), 265–270.

Bagheri, S., Mohd Hir, Z. A., Termeh Yousefi, A., & Abd Hamid, S. B. (2015c). Photocatalytic performance of activated carbon-supported mesoporous titanium dioxide. *Desalination and Water Treatment*(ahead-of-print), 1–7.

Bagheri, S., Muhd Julkapli, N., & Bee Abd Hamid, S. (2014d). Titanium Dioxide as a Catalyst Support in Heterogeneous Catalysis. *The Scientific World Journal, 2014*.

Bagheri, S., Ramimoghadam, D., Yousefi, A. T., & Hamid, S. B. A. (2015d). Synthesis, Characterization and Electrocatalytic Activity of Silver Doped-Titanium Dioxide Nanoparticles. *Int. J. Electrochem. Sci, 10*, 3088–3097.

Bagheri, S., Shameli, K., & Abd Hamid, S. B. (2012b). Synthesis and characterization of anatase titanium dioxide nanoparticles using egg white solution via Sol-Gel method. *Journal of Chemistry, 2013*.

Bagheri, S., & Ziglari, A. (2015). Density functional theory and QM/MM illustration of the behavior of B23N23 nano-cone: EPR & NMR investigation. *Oriental Journal of Chemistry, 31*(2), 857–866.

Bai, Y., Sun, Y., & Sun, C. (2008). Pt–Pb nanowire array electrode for enzyme-free glucose detection. *Biosensors and Bioelectronics, 24*(4), 579–585. doi: 10.1016/j.bios.2008.06.003

Bakker, E. (2004). Electrochemical sensors. *Analytical Chemistry-Washington DC-, 76*, 3285–3298.

Balasubramanian, K., & Burghard, M. (2006). Biosensors based on carbon nanotubes. *Analytical and bioanalytical chemistry, 385*(3), 452–468.

Bard, A. J., & Faulkner, L. R. (1980). *Electrochemical methods: fundamentals and applications* (Vol. 2): Wiley New York.

Barrera, C., Zhukov, I., Villagra, E., Bedioui, F., Páez, M. A., Costamagna, J., & Zagal, J. H. (2006). Trends in reactivity of unsubstituted and substituted cobalt-phthalocyanines for the electrocatalysis of glucose oxidation. *Journal of Electroanalytical Chemistry, 589*(2), 212–218. doi: 10.1016/j.jelechem.2006.02.009

Belton, D., Paine, G., Patwardhan, S. V., & Perry, C. C. (2004). Towards an understanding of (bio)silicification: the role of amino acids and lysine oligomers in silicification. [10.1039/B401882F]. *Journal of Materials Chemistry, 14*(14), 2231–2241. doi: 10.1039/b401882f

Berchmans, S., Gomathi, H., & Rao, G. P. (1995). Electrooxidation of alcohols and sugars catalysed on a nickel oxide modified glassy carbon electrode. *Journal of Electroanalytical Chemistry, 394*(1–2), 267–270. doi: 10.1016/0022-0728(95)04099-a

Besada, A. (1987). A New Simple and Sensitive Spectrophotometric Procedure for Determination of Adrenaline. *Analytical letters, 20*(3), 427–434. doi: 10.1080/00032718708064578

Bhadra, P., Mitra, M. K., Das, G. C., Dey, R., & Mukherjee, S. (2011). Interaction of chitosan capped ZnO nanorods with Escherichia coli. *Materials Science and Engineering: C, 31*(5), 929–937. doi: http://dx.doi.org/10.1016/j.msec.2011.02.015

Bhat, A. A., & Pangarkar, V. G. (2000). Methanol-selective membranes for the pervaporative separation of methanol–toluene mixtures. *Journal of Membrane Science, 167*(2), 187–201. doi: 10.1016/s0376-7388(99)00289-6

Bickley, R. I., Gonzalez-Carreno, T., Lees, J. S., Palmisano, L., & Tilley, R. J. D. (1991). A structural investigation of titanium dioxide photocatalysts. *Journal of Solid State Chemistry, 92*(1), 178–190.

Bisenberger, M., Bräuchle, C., & Hampp, N. (1995). A triple-step potential waveform at enzyme multisensors with thick-film gold electrodes for detection of glucose and sucrose. *Sensors and Actuators B: Chemical, 28*(3), 181–189. doi: 10.1016/0925-4005(95)01730-5

Bowers, L. D., & Carr, P. W. (1976). Applications of immobilized enzymes in analytical chemistry. *Analytical Chemistry, 48*(7), 544A–559a. doi: 10.1021/ac60371a033

Brett, C. M. A., & Brett, A. M. O. (1993). *Electrochemistry: principles, methods, and applications* (Vol. 4): Oxford University Press Oxford.

Brinker, C. J., & Scherer, G. W. (1990). *Sol-gel science: the physics and chemistry of sol-gel processing*: Academic Pr.

Buratti, S., Brunetti, B., & Mannino, S. (2008). Amperometric detection of carbohydrates and thiols by using a glassy carbon electrode coated with Co oxide/multi-wall carbon nanotubes catalytic system. *Talanta, 76*(2), 454–457. doi: 10.1016/j.talanta.2008.03.031

Burda, C., Chen, X., Narayanan, R., & El-Sayed, M. A. (2005). Chemistry and Properties of Nanocrystals of Different Shapes. *Chemical Reviews, 105*(4), 1025–1102. doi: 10.1021/cr030063a

C. N. Likos, K. A. V., H. Lwen, & N. J. Wagner. (2009). Colloidal Stabilization by Adsorbed Gelatin. *langmuir, 16*(9), 4100–4108. doi: 10.1021/la991142d

Cai, H., Xu, Y., Zhu, N., He, P., & Fang, Y. (2002). An electrochemical DNA hybridization detection assay based on a silver nanoparticle label. *Analyst, 127*(6), 803–808.

Campbell, L. K., Na, B. K., & Ko, E. I. (1992). Synthesis and characterization of titania aerogels. *Chemistry of Materials, 4*(6), 1329–1333. doi: 10.1021/cm00024a037

Carp, O., Huisman, C. L., & Reller, A. (2004). Photoinduced reactivity of titanium dioxide. *Progress in Solid State Chemistry, 32*(1), 33–177.

Carreño-Gómez, B., & Duncan, R. (1997). Evaluation of the biological properties of soluble chitosan and chitosan microspheres. *International Journal of Pharmaceutics, 148*(2), 231–240. doi: 10.1016/s0378-5173(96)04847-8

Cavrini, V., Gatti, R., Roveri, P., & Cesaroni, M. R. (1988). Use of 4-(6-methylnaphthalen-2-yl)-4-oxobut-2-enoic acid as a reagent for the spectrophotometric and fluorimetric determination of aliphatic thiol drugs. *Analyst, 113*(9), 1447–1452.

Chang, B. Y., & Park, S. M. (2010). Electrochemical impedance spectroscopy. *Annual Review of Analytical Chemistry, 3*, 207–229.

Chaturvedi, S., Dave, P. N., & Shah, N. K. (2012). Applications of nano-catalyst in new era. *Journal of Saudi Chemical Society, 16*(3), 307–325. doi: 10.1016/j.jscs.2011.01.015

Chekin, F., Bagheri, S., & Abd Hamid, S. B. (2014a). Green synthesis of Ag nanoparticles by Callicarpa Maingayi: Characterization and its application with graphene oxide for enzymeless hydrogen peroxide detection. *Journal of the Chinese Chemical Society, 61*(6), 631–637.

Chekin, F., Bagheri, S., & Abd Hamid, S. B. (2015a). Functionalization of Graphene Oxide with 3-Mercaptopropyltrimethoxysilane and Its Electrocatalytic Activity in Aqueous Medium. *Journal of the Chinese Chemical Society, 62*(8), 689–694.

Chekin, F., Bagheri, S., Arof, A. K., & Hamid, S. B. A. (2012a). Preparation and characterization of Ni (II)/polyacrylonitrile and carbon nanotube composite modified electrode and application for carbohydrates electrocatalytic oxidation. *Journal of Solid State Electrochemistry, 16*(10), 3245–3251.

Chekin, F., Bagheri, S., & Hamid, S. A. (2012b). Electrochemistry and electrocatalysis of cobalt (II) immobilized onto gel-assisted synthesized zinc oxide nanoparticle–multi wall carbon nanotube–polycaprolactone composite film: application to determination of glucose. *Analytical Methods, 4*(8), 2423–2428.

Chekin, F., Bagheri, S., & Hamid, S. B. A. (2013a). Synthesis of Pt doped TiO_2 nanoparticles: characterization and application for electrocatalytic oxidation of l-methionine. *Sensors and Actuators B: Chemical, 177*, 898–903.

Chekin, F., Bagheri, S., & Hamid, S. B. A. (2014b). Gel-assisted synthesis of Ag nanoparticles: a novel hydrogen peroxide sensor based on Ag nanoparticles-carbon nanotube composite film. *Russian Journal of Electrochemistry, 50*(12), 1164–1169.

Chekin, F., Bagheri, S., & Hamid, S. B. A. (2014c). Preparation and electrochemical performance of graphene–Pt black nanocomposite for electrochemical methanol oxidation. *Journal of Solid State Electrochemistry, 18*(4), 893–898.

Chekin, F., Bagheri, S., & Hamid, S. B. A. (2014d). Synthesis of graphene oxide nanosheet: A novel glucose sensor based on nickel-graphene oxide composite film. *Russian Journal of Electrochemistry, 50*(11), 1044–1049.

Chekin, F., Bagheri, S., & Hamid, S. B. A. (2015b). Synthesis and spectroscopic characterization of palladium-doped titanium dioxide catalyst. *Bull. Mater. Sci, 38*(2).

Chekin, F., Bagherib, S., & Hamidb, S. B. A. (2013b). Synthesis of Tungsten Oxide Nanorods by the Controlling Precipitation Reaction: Application for Hydrogen Evolution Reaction on a WO3 Nanorods/Carbon Nanotubes Composite Film Modified Electrode. *J. Chin. Chem. Soc, 60*, 447–451.

Chekin, F., Raoof, J., Bagheri, S., & Hamid, S. A. (2012c). The porous chitosan–sodium dodecyl sulfate–carbon nanotube nanocomposite: direct electrochemistry and electrocatalysis of hemoglobin. *Analytical Methods, 4*(9), 2977–2981.

Chekin, F., Raoof, J. B., Bagheri, S., & Hamid, S. B. A. (2012d). Fabrication of chitosan-multiwall carbon nanotube nanocomposite containing ferri/ferrocyanide: application for simultaneous detection of D-penicillamine and tryptophan. *Journal of the Chinese Chemical Society, 59*(11), 1461–1467.

Chen, Q., Wang, J., Rayson, G., Tian, B., & Lin, Y. (1993). Sensor array for carbohydrates and amino acids based on electrocatalytic modified electrodes. *Analytical Chemistry, 65*(3), 251–254. doi: 10.1021/ac00051a011

Chen, S. M., & Chzo, W. Y. (2006). Simultaneous voltammetric detection of dopamine and ascorbic acid using didodecyldimethylammonium bromide (DDAB) film-modified electrodes. *Journal of Electroanalytical Chemistry, 587*(2), 226–234.

Chen, X., Pan, H., Liu, H., & Du, M. (2010). Nonenzymatic glucose sensor based on flower-shaped Au@Pd core–shell nanoparticles–ionic liquids composite film modified glassy carbon electrodes. *Electrochimica Acta, 56*(2), 636–643. doi: 10.1016/j.electacta.2010.10.001

Cheng, B., & Samulski, E. T. (2004). Hydrothermal synthesis of one-dimensional ZnO nanostructures with different aspect ratios. [10.1039/B316435G]. *Chemical Communications*(8), 986–987. doi: 10.1039/b316435g

Cheng, H., Cheng, J., Zhang, Y., & Wang, Q.-M. (2007). Large-scale fabrication of ZnO micro-and nano-structures by microwave thermal evaporation deposition. *Journal of Crystal Growth, 299*(1), 34–40. doi: 10.1016/j.jcrysgro.2006.10.245

Cho, S., Jung, S. H., & Lee, K. H. (2008). Morphology-Controlled Growth of ZnO Nanostructures Using Microwave Irradiation: from Basic to Complex Structures. *The Journal of Physical Chemistry C, 112*(33), 12769–12776. doi: 10.1021/jp803783s

Corradi, A. B., Bondioli, F., Focher, B., Ferrari, A. M., Grippo, C., Mariani, E., & Villa, C. (2005). Conventional and Microwave-Hydrothermal Synthesis of TiO$_2$ Nanopowders. *Journal of the American Ceramic Society, 88*(9), 2639–2641.

Covington, A. K. (1979). *Ion-selective electrode methodology*: CRC press.

Cox, D. M., Trevor, D. J., Whetten, R. L., & Kaldor, A. (1988). Aluminum clusters: ionization thresholds and reactivity toward deuterium, water, oxygen, methanol, methane, and carbon monoxide. *The Journal of Physical Chemistry, 92*(2), 421–429. doi: 10.1021/j100313a036

D'Eramo, F., Marioli, J. M., Arévalo, A. A., & Sereno, L. E. (1999). HPLC Analysis of Carbohydrates with Electrochemical Detection at a Poly-1-naphthylamine/Copper Modified Electrode. *Electroanalysis, 11*(7), 481–486. doi: 10.1002/(sici)1521-4109 (199906)11:7<481::aid-elan481>3.0.co;2-7

D'Eramo, F., Marioli, J. M., Arévalo, A. H., & Sereno, L. E. (2003). Optimization of the electrodeposition of copper on poly-1-naphthylamine for the amperometric detection of carbohydrates in HPLC. *Talanta, 61*(3), 341–352. doi: 10.1016/s0039-9140(03)00304-7

Daniel, M.-C., & Astruc, D. (2003). Gold Nanoparticles: Assembly, Supramolecular Chemistry, Quantum-Size-Related Properties, and Applications toward Biology, Catalysis, and Nanotechnology. *Chemical Reviews, 104*(1), 293–346. doi: 10.1021/cr030698+

de Mele, M. F. L., Videla, H. A., & Arvía, A. J. (1983). The electrooxidation of glucose on platinum electrodes in buffered media. *Bioelectrochemistry and Bioenergetics, 10*(2–3), 239–249. doi: 10.1016/0302-4598(83)85082-x

De Wael, K., De Belder, S., Van Vlierberghe, S., Van Steenberge, G., Dubruel, P., & Adriaens, A. (2010). Electrochemical study of gelatin as a matrix for the immobilization of horse heart cytochrome c. *Talanta, 82*(5), 1980–1985. doi: 10.1016/j.talanta.2010.08.019

Dickerson, M. B., Naik, R. R., Stone, M. O., Cai, Y., & Sandhage, K. H. (2004). Identification of peptides that promote the rapid precipitation of germania nanoparticle networks via use of a peptide display library. [10.1039/B402480J]. *Chemical Communications*(15), 1776–1777. doi: 10.1039/b402480j

Diebold, U. (2003). The surface science of titanium dioxide. *Surface science reports, 48*(5), 53–229.

Ding, Z., Hu, X., Yue, P. L., Lu, G. Q., & Greenfield, P. F. (2001). Synthesis of anatase TiO$_2$ supported on porous solids by chemical vapor deposition. *Catalysis Today, 68*(1), 173–182.

Dong, S., & Wang, Y. (1989). The application of chemically modified electrodes in analytical chemistry. *Electroanalysis, 1*(2), 99–106.

Dong, X.-C., Xu, H., Wang, X.-W., Huang, Y.-X., Chan-Park, M. B., Zhang, H., . . . Chen, P. (2012). 3D Graphene–Cobalt Oxide Electrode for High-Performance Supercapacitor and Enzymeless Glucose Detection. *ACS Nano, 6*(4), 3206–3213. doi: 10.1021/nn300097q

Donya Ramimoghadam, S. B., Amin Termeh Yousefi, Sharifah Bee Abd Hamid. (2015). Statistical optimization of effective parameters on saturation magnetization of nanomagnetite particles. *Journal of Magnetism and Magnetic Materials.*

Eftekhari, A., Molaei, F., & Arami, H. (2006). Flower-like bundles of ZnO nanosheets as an intermediate between hollow nanosphere and nanoparticles. *Mater. Sci. Eng. A, 437*(2), 446–450. doi: 10.1016/j.msea.2006.08.033

Elen, K., Van den Rul, H., Hardy, A., Van Bael, M. K., D'Haen, J., Peeters, R., ... Mullens, J. (2009). Hydrothermal synthesis of ZnO nanorods: a statistical determination of the significant parameters in view of reducing the diameter. *Nanotechnology, 20*(5), 055608.

Ellmer, K., Klein, A., & Rech, B. (2008). *Transparent Conductive Zinc Oxide: Basics and Applications in Thin Film Solar Cells* (Vol. 104): Springer.

Espitia, P. J. P., Soares, N. F. F., Coimbra, J. S. R., de Andrade, N. J., Cruz, R. S., & Medeiros, E. A. A. (2012). Zinc Oxide Nanoparticles: Synthesis, Antimicrobial Activity and Food Packaging Applications. *Food and Bioprocess Technology*, 1–18.

Fan, Z., & Lu, J. G. (2005). Zinc Oxide Nanostructures: Synthesis and Properties. *J. Nanosci. Nanotechnol., 5*(10), 1561–1573. doi: 10.1166/jnn.2005.182

Fang, B., Wei, Y., Li, M., Wang, G., & Zhang, W. (2007). Study on electrochemical behavior of tryptophan at a glassy carbon electrode modified with multi-walled carbon nanotubes embedded cerium hexacyanoferrate. *Talanta, 72*(4), 1302–1306. doi: 10.1016/j.talanta.2007.01.039

Farley, N. R. S., Staddon, C. R., Zhao, L., Edmonds, K. W., Gallagher, B. L., & Gregory, D. H. (2004). Sol-gel formation of ordered nanostructured doped ZnO films. *Journal of Materials Chemistry, 14*(7), 1087–1092.

Fernández-Sánchez, C., McNeil, C. J., & Rawson, K. (2005). Electrochemical impedance spectroscopy studies of polymer degradation: application to biosensor development. *TrAC Trends in Analytical Chemistry, 24*(1), 37–48.

Ferraro, J. R., & Basile, L. J. (1975). Fourier transform infrared spectroscopy: applications to chemical systems. Volume 4.

Fiorito, P. A., Gonçales, V. R., Ponzio, E. A., & de Torresi, S. I. C. (2004). Synthesis, characterization and immobilization of Prussian blue nanoparticles. A potential tool for biosensing devices. *Chem. Commun.*(3), 366–368.

Fiorucci, A. R., & Cavalheiro, É. T. G. (2002). The use of carbon paste electrode in the direct voltammetric determination of tryptophan in pharmaceutical formulations. *Journal of Pharmaceutical and Biomedical Analysis, 28*(5), 909–915. doi: 10.1016/s0731-7085(01)00711-7

Fleischmann, M., Korinek, K., & Pletcher, D. (1971). The oxidation of organic compounds at a nickel anode in alkaline solution. *Journal of Electroanalytical Chemistry and Interfacial Electrochemistry, 31*(1), 39–49. doi: 10.1016/s0022-0728(71)80040-2

Fleischmann, M., Korinek, K., & Pletcher, D. (1972). The oxidation of hydrazine at a nickel anode in alkaline solution. *Journal of Electroanalytical Chemistry and Interfacial Electrochemistry, 34*(2), 499–503. doi: http://dx.doi.org/10.1016/S0022-0728(72)80425-X

Frith, K. A., & Limson, J. L. (2009). pH tuning of Nafion® for selective detection of tryptophan. *Electrochimica Acta, 54*(13), 3600–3605. doi: 10.1016/j.electacta.2009.01.028

Fu, C., Yang, W., Chen, X., & Evans, D. G. (2009). Direct electrochemistry of glucose oxidase on a graphite nanosheet–Nafion composite film modified electrode. *Electrochemistry Communications, 11*(5), 997–1000. doi: 10.1016/j.elecom.2009.02.042

Fujishima, A., Rao, T. N., & Tryk, D. A. (2000). Titanium dioxide photocatalysis. *Journal of Photochemistry and Photobiology C: Photochemistry Reviews, 1*(1), 1–21.

German, N., Ramanavicius, A., Voronovic, J., & Ramanaviciene, A. Glucose biosensor based on glucose oxidase and gold nanoparticles of different sizes covered by polypyrrole layer. *Colloids and Surfaces A: Physicochemical and Engineering Aspects*(0). doi: 10.1016/j.colsurfa.2012.02.012

Gholami, T., Bazarganipour, M., Salavati-Niasari, M., & Bagheri, S. Photocatalytic degradation of methylene blue on TiO_2@ SiO_2 core/shell nanoparticles: synthesis and characterization. *Journal of Materials Science: Materials in Electronics*, 1–8.

Gholamrezaei, S., Salavati-Niasari, M., Bazarganipour, M., Panahi-Kalamuei, M., & Bagheri, S. (2014). Novel precursors for synthesis of dendrite-like PbTe nanostructures and investigation of photoluminescence behavior. *Advanced Powder Technology, 25*(5), 1585–1592.

Gholivand, M. B., Pashabadi, A., Azadbakht, A., & Menati, S. (2011). A nano-structured Ni(II)–ACDA modified gold nanoparticle self-assembled electrode for electrocatalytic oxidation and determination of tryptophan. *Electrochimica Acta, 56*(11), 4022–4030. doi: 10.1016/j.electacta.2011.02.009

Giusti, M. M., & Wrolstad, R. E. (2001). Characterization and measurement of anthocyanins by UV-visible spectroscopy. *Current protocols in food analytical chemistry*.

Gojny, F. H., Nastalczyk, J., Roslaniec, Z., & Schulte, K. (2003). Surface modified multi-walled carbon nanotubes in CNT/epoxy-composites. *Chemical Physics Letters, 370*(5), 820–824.

Goldstein, J., Newbury, D. E., Joy, D. C., Lyman, C. E., Echlin, P., Lifshin, E., . . . Michael, J. R. (2003). *Scanning electron microscopy and X-ray microanalysis*: Springer.

Gómez, R., López, T., Ortiz-Islas, E., Navarrete, J., Sánchez, E., Tzompantzi, F., & Bokhimi, X. (2003). Effect of sulfation on the photoactivity of TiO$_2$ sol-gel derived catalysts. *Journal of Molecular Catalysis A: Chemical, 193*(1–2), 217–226. doi: 10.1016/s1381-1169(02)00473-9

Gong, K., Zhang, M., Yan, Y., Su, L., Mao, L., Xiong, S., & Chen, Y. (2004). Sol-Gel-Derived Ceramic–Carbon Nanotube Nanocomposite Electrodes:Tunable Electrode Dimension and Potential Electrochemical Applications. *Analytical Chemistry, 76*(21), 6500–6505. doi: 10.1021/ac0492867

Gopalan, A. I., Lee, K. P., Ragupathy, D., Lee, S. H., & Lee, J. W. (2009). An electrochemical glucose biosensor exploiting a polyaniline grafted multiwalled carbon nanotube/ perfluorosulfonate ionomer–silica nanocomposite. *Biomaterials, 30*(30), 5999–6005. doi: 10.1016/j.biomaterials.2009.07.047

Gosser, D. K. (1993). *Cyclic voltammetry: simulation and analysis of reaction mechanisms*: VCH New York.

Goyal, R. N., Bishnoi, S., & Agrawal, B. (2011). Single-Walled-Carbon-Nanotube-Modified Pyrolytic Graphite Electrode Used as a Simple Sensor for the Determination of Salbutamol in Urine. *International Journal of Electrochemistry, 2011*. doi: 10.4061/2011/373498

Goyal, R. N., Bishnoi, S., Chasta, H., Aziz, M. A., & Oyama, M. (2011). Effect of surface modification of indium tin oxide by nanoparticles on the electrochemical determination of tryptophan. *Talanta, 85*(5), 2626–2631. doi: 10.1016/j.talanta.2011.08.031

Grieshaber, D., MacKenzie, R., Voeroes, J., & Reimhult, E. (2008). Electrochemical biosensors-Sensor principles and architectures. *Sensors, 8*(3), 1400–1458.

Griffiths, P., & De Haseth, J. A. (2007). *Fourier transform infrared spectrometry* (Vol. 171): Wiley-Interscience.

Gülce, H., Çelebi, S. S., Özyörük, H., & Yildiz, A. (1995). Amperometric enzyme electrode for sucrose determination prepared from glucose oxidase and invertase co-immobilized in poly(vinylferrocenium). *Journal of Electroanalytical Chemistry, 397*(1–2), 217–223. doi: http://dx.doi.org/10.1016/0022-0728(95)04192-1

Hampson, N. A., Lee, J. B., & Macdonald, K. I. (1972). Oxidations at copper electrodes: Part 41. The oxidation of α-amino acids. *Journal of Electroanalytical Chemistry and Interfacial Electrochemistry, 34*(1), 91–99. doi: http://dx.doi.org/10.1016/S0022-0728(72)80505-9

Heineman, W. R., Wieck, H. J., & Yacynych, A. M. (1980). Polymer film chemically modified electrode as a potentiometric sensor. *Analytical Chemistry, 52*(2), 345–346.

Heli, H., Majdi, S., & Sattarahmady, N. (2010). Ultrasensitive sensing of N-acetyl-l-cysteine using an electrocatalytic transducer of nanoparticles of iron(III) oxide core–cobalt hexacyanoferrate shell. *Sensors and Actuators B: Chemical, 145*(1), 185–193. doi: 10.1016/j.snb.2009.11.065

Heller, A., & Yarnitzky, C. (2001). Potentiometric sensors for analytic determination: Google Patents.

Hench, L. L., & Ulrich, D. R. (1986). *Science of ceramic chemical processing*: Wiley-Interscience.

Hernandez-Velez, M. (2006). Nanowires and 1D arrays fabrication: An overview. *Thin solid films, 495*(1), 51–63.

Heyrovský, J. (1956). The development of polarographic analysis. *Analyst, 81*(961), 189–192.

Hoffmann, M. R., Martin, S. T., Choi, W., & Bahnemann, D. W. (1995). Environmental Applications of Semiconductor Photocatalysis. *Chemical Reviews, 95*(1), 69–96. doi: 10.1021/cr00033a004

Huang, K.-J., Xu, C.-X., Xie, W.-Z., & Wang, W. (2009). Electrochemical behavior and voltammetric determination of tryptophan based on 4-aminobenzoic acid polymer film modified glassy carbon electrode. *Colloids and Surfaces B: Biointerfaces, 74*(1), 167–171. doi: 10.1016/j.colsurfb.2009.07.013

Huang, W., Hu, W., & Song, J. (2003). Adsorptive stripping voltammetric determination of 4-aminophenol at a single-wall carbon nanotubes film coated electrode. *Talanta, 61*(3), 411–416.

Huang, W., Mai, G., Liu, Y., Yang, C., & Qua, W. (2004). Voltammetric determination of tryptophan at a single-wall carbon nanotubes modified electrode. *Journal of Nanoscience and Nanotechnology, 4*(4), 423–427.

Hubalek, J., Hradecky, J., Adam, V., Krystofova, O., Huska, D., Masarik, M., . . . Adamek, M. (2007). Spectrometric and voltammetric analysis of urease–Nickel nanoelectrode as an electrochemical sensor. *Sensors, 7*(7), 1238–1255.

Hutton, L. A., Vidotti, M., Patel, A. N., Newton, M. E., Unwin, P. R., & Macpherson, J. V. (2010). Electrodeposition of Nickel Hydroxide Nanoparticles on Boron-Doped Diamond Electrodes for Oxidative Electrocatalysis. *The Journal of Physical Chemistry C, 115*(5), 1649–1658. doi: 10.1021/jp109526b

Iijima, S. (1991). Helical microtubules of graphitic carbon. *Nature, 354*(6348), 56–58.

Jalal, R., Goharshadi, E. K., Abareshi, M., Moosavi, M., Yousefi, A., & Nancarrow, P. (2010). ZnO nanofluids: green synthesis, characterization, and antibacterial activity. *Materials Chemistry and Physics, 121*(1), 198–201.

Janata, J. (2002). Electrochemical Sensors and their impedances: a tutorial. *Critical reviews in analytical chemistry, 32*(2), 109–120.

Janata, J. (2009). *Principles of chemical sensors*: Springer.

Janata, J., & Bezegh, A. (1988). Chemical sensors. *Analytical Chemistry, 60*(12), 62R–74R. doi: 10.1021/ac00163a004

Janata, J., & Huber, R. J. (1985). *Solid state chemical sensors*: Academic Press Orlando.

Janek, R. P., Fawcett, W. R., & Ulman, A. (1998). Impedance spectroscopy of self-assembled monolayers on Au (111): sodium ferrocyanide charge transfer at modified electrodes. *Langmuir, 14*(11), 3011–3018.

Jiang, L., Liu, C., Peng, Z., & Lu, G. (2004). A chitosan-multiwall carbon nanotube modified electrode for simultaneous detection of dopamine and ascorbic acid. *Analytical sciences, 20*(7), 1055–1059.

Jiang, L., Wang, R., Li, X., & Lu, G. (2005). Electrochemical oxidation behavior of nitrite on a chitosan-carboxylated multiwall carbon nanotube modified electrode. *Electrochemistry Communications, 7*(6), 597–601.

Jiang, L. Y., Liu, C. Y., Jiang, L. P., & Lu, G. H. (2005). A multiwall carbon nanotube-chitosan modified electrode for selective detection of dopamine in the presence of ascorbic acid. *Chinese Chemical Letters, 16*(2), 229–232.

Jin, W., Lee, I., Kompch, A., Dorfler, U., & Wintere, M. (2007). Chemical vapor synthesis and characterization of chromium doped zinc oxide nanoparticles. *Journal of the European Ceramic Society., 27*(13–15), 4333–4337. doi: 10.1016/j.jeurceramsoc.2007.02.152

Julkapli, N. M., & Bagheri, S. (2015). Graphene supported heterogeneous catalysts: an overview. International Journal of Hydrogen Energy, 40(2), 948–979.

Julkapli, N. M., Bagheri, S., & Sapuan, S. (2015a). Bio-nanocomposites from Natural Fibre Derivatives: Manufacturing and Properties. In Manufacturing of Natural Fibre Reinforced Polymer Composites (pp. 233–265): Springer.

Julkapli, N. M., Bagheri, S., & Sapuan, S. (2015b). Multifunctionalized Carbon Nanotubes Polymer Composites: Properties and Applications. In Eco-friendly Polymer Nanocomposites (pp. 155–214): Springer.

Kalcher, K., Kauffmann, J. M., Wang, J., Švancara, I., Vytřas, K., Neuhold, C., & Yang, Z. (2005). Sensors based on carbon paste in electrochemical analysis: a review with particular emphasis on the period 1990–1993. *Electroanalysis, 7*(1), 5–22.

Karakitsou, K. E., & Verykios, X. E. (1993). Effects of altervalent cation doping of titania on its performance as a photocatalyst for water cleavage. *The Journal of Physical Chemistry, 97*(6), 1184–1189. doi: 10.1021/j100108a014

Kargar, M., Alikhanzadeh-Arani, S., Bagheri, S., & Salavati-Niasari, M. (2014). Magnetic and structural characteristics of HoBa 2 Cu 3 O 7-x nanorods synthesized in the presence of an appropriate surfactant. Ceramics International, 40(7), 11109–11114.

Kargar, M., Alikhanzadeh-Arani, S., Salavati-Niasari, M., & Bagheri, S. (2015). Characterization of REBa 2 Cu 3 O 7-X (RE=Gd, Ho) nanostructures, fabricated by a simple technique. Physica C: Superconductivity and its Applications, 511, 20–25.

Katz, E., & Willner, I. (2003). Probing biomolecular interactions at conductive and semi-conductive surfaces by impedance spectroscopy: routes to impedimetric immunosensors, DNA-sensors, and enzyme biosensors. *Electroanalysis, 15*(11), 913–947.

Kawanishi, N., Christenson, H. K., & Ninham, B. W. (1990). Measurement of the interaction between adsorbed polyelectrolytes: gelatin on mica surfaces. *The Journal of Physical Chemistry, 94*(11), 4611–4617. doi: 10.1021/j100374a045

Kim, I. J., Han, S. D., Han, C. H., Gwak, J., Lee, H. D., & Wang, J. S. (2006). Micro semiconductor CO sensors based on indium-doped tin dioxide nanocrystalline powders. *Sensors, 6*(5), 526–535.

Kim, K. D., Choi, D. W., Choa, Y. H., & Hee, T. K. (2007). Optimization of parameters for the synthesis of zinc oxide nanoparticles by Taguchi robust design method. *Colloids and Surfaces A: Physicochemical and Engineering Aspects, 311*(1–3), 170–173. doi: 10.1016/j.colsurfa.2007.06.017

Kim, K. D., Lee, T. J., & Kim, H. T. (2003). Optimal conditions for synthesis of TiO_2 nanoparticles in semi-batch reactor. *Colloids and Surfaces A: Physicochemical and Engineering Aspects, 224*(1–3), 1–9. doi: 10.1016/s0927-7757(03)00256-5

Kim, K. H., & Jo, W. H. (2007). Synthesis of Polythiophene-graft-PMMA and Its Role as Compatibilizer for Poly(styrene-co-acrylonitrile)/MWCNT Nanocomposites. *Macromolecules, 40*(10), 3708–3713. doi: 10.1021/ma070127+

Kim, P., & Lieber, C. M. (1999). Nanotube Nanotweezers. *Science, 286*(5447), 2148–2150. doi: 10.1126/science.286.5447.2148

Klein, L. C. (1994). *Sol-gel optics: Processing and applications*: Springer.

Klug, H. P., & Alexander, L. E. (1974). X-ray diffraction procedures: for polycrystalline and amorphous materials. *X-Ray Diffraction Procedures: For Polycrystalline and Amorphous Materials, 2nd Edition, by Harold P. Klug, Leroy E. Alexander, pp. 992. ISBN 0-471-49369-4. Wiley-VCH, May 1974.*, 1.

Kriz, D., Kempe, M., & Mosbach, K. (1996). Introduction of molecularly imprinted polymers as recognition elements in conductometric chemical sensors. *Sensors and Actuators B: Chemical, 33*(1), 178–181.

Kroto, H. W., Heath, J. R., O'Brien, S. C., Curl, R. F., & Smalley, R. E. (1985). C 60: buckminsterfullerene. *Nature, 318*(6042), 162–163.

Kuo, C. L., Kuo, T. J., & Huang, M. H. (2005). Hydrothermal synthesis of ZnO microspheres and hexagonal microrods with sheetlike and platelike nanostructures. *The Journal of Physical Chemistry B, 109*(43), 20115–20121.

Lakshmi, B. B., Dorhout, P. K., & Martin, C. R. (1997). Sol-gel template synthesis of semiconductor nanostructures. *Chemistry of Materials, 9*(3), 857–862.

Laviron, E. (1979). The use of linear potential sweep voltammetry and of a.c. voltammetry for the study of the surface electrochemical reaction of strongly adsorbed systems and of redox modified electrodes. *Journal of Electroanalytical Chemistry and Interfacial Electrochemistry, 100*(1–2), 263–270. doi: http://dx.doi.org/10.1016/S0022-0728(79)80167-9

Lee, J. U., Huh, J., Kim, K. H., Park, C., & Jo, W. H. (2007). Aqueous suspension of carbon nanotubes via non-covalent functionalization with oligothiophene-terminated poly(ethylene glycol). *Carbon, 45*(5), 1051–1057. doi: 10.1016/j.carbon.2006.12.017

Lesho, M. J., & Sheppard, N. F. (1996). Adhesion of polymer films to oxidized silicon and its effect on performance of a conductometric pH sensor. *Sensors and Actuators B: Chemical, 37*(1), 61–66.

Leutwyler, W. K., Bürgi, S. L., & Burgl, H. (1996). Semiconductor clusters, nanocrystals, and quantum dots. *Science, 271*, 933.

Li, G., Li, L., Boerio-Goates, J., & Woodfield, B. F. (2005). High purity anatase TiO_2 nanocrystals: near room-temperature synthesis, grain growth kinetics, and surface hydration chemistry. *Journal of the American Chemical Society, 127*(24), 8659–8666.

Li, Q., Zaiser, M., & Koutsos, V. (2004). Carbon nanotube/epoxy resin composites using a block copolymer as a dispersing agent. *Physica status solidi (A), 201*(13), R89–R91.

Li, S., He, P., Dong, J., Guo, Z., & Dai, L. (2005). DNA-directed self-assembling of carbon nanotubes. *Journal of the American Chemical Society, 127*(1), 14–15.

Liang, W., & Zhuobin, Y. (2003). Direct electrochemistry of glucose oxidase at a gold electrode modified with single-wall carbon nanotubes. *Sensors, 3*(12), 544–554.

Liang, Y., Gan, S., Chambers, S. A., & Altman, E. I. (2001). Surface structure of anatase TiO_2 (001): Reconstruction, atomic steps, and domains. *Physical Review B, 63*(23), 235402.

Lim, S. H., Wei, J., Lin, J., Li, Q., & KuaYou, J. (2005). A glucose biosensor based on electrodeposition of palladium nanoparticles and glucose oxidase onto Nafion-solubilized carbon nanotube electrode. *Biosensors and Bioelectronics, 20*(11), 2341–2346.

Lin, C.-C., & Yang, M.-C. (2003). Cholesterol Oxidation Using Hollow Fiber Dialyzer Immobilized with Cholesterol Oxidase: Preparation and Properties. *Biotechnology Progress, 19*(2), 361–364. doi: 10.1021/bp0256014

Lin, C. C., & Li, Y. Y. (2009). Synthesis of ZnO nanowires by thermal decomposition of zinc acetate dihydrate. *Materials Chemistry and Physics, 113*(1), 334–337.

Lin, Y., Lu, F., Tu, Y., & Ren, Z. (2003). Glucose Biosensors Based on Carbon Nanotube Nanoelectrode Ensembles. *Nano Letters, 4*(2), 191–195. doi: 10.1021/nl0347233

Lin, Y., Yantasee, W., & Wang, J. (2005). Carbon nanotubes (CNTs) for the development of electrochemical biosensors. *Frontiers in Bioscience, 10*(1), 492–505.

Linford, R. G. (1990). *Electrochemical science and technology of polymers* (Vol. 2): Kluwer Academic Pub.

Liu, B., & Zeng, H. C. (2003). Hydrothermal Synthesis of ZnO Nanorods in the Diameter Regime of 50 nm. *J. Am. Chem. Soc., 125*(15), 4430–4431. doi: 10.1021/ja0299452

Liu, G., & Lin, Y. (2006). Amperometric glucose biosensor based on self-assembling glucose oxidase on carbon nanotubes. *Electrochemistry Communications, 8*(2), 251–256. doi: 10.1016/j.elecom.2005.11.015

Liu, X., Luo, L., Ding, Y., & Ye, D. (2011). Poly-glutamic acid modified carbon nanotube-doped carbon paste electrode for sensitive detection of L-tryptophan. *Bioelectrochemistry, 82*(1), 38–45. doi: 10.1016/j.bioelechem.2011.05.001

Liu, Y., Tang, J., Chen, X., & Xin, J. H. (2005). Decoration of carbon nanotubes with chitosan. *Carbon, 43*(15), 3178–3180. doi: 10.1016/j.carbon.2005.06.020

Liu, Y., Wang, M., Zhao, F., Xu, Z., & Dong, S. (2005). The direct electron transfer of glucose oxidase and glucose biosensor based on carbon nanotubes/chitosan matrix. *Biosensors and Bioelectronics, 21*(6), 984–988.

Liu, Y., Yuan, R., Chai, Y., Tang, D., Dai, J., & Zhong, X. (2006). Direct electrochemistry of horseradish peroxidase immobilized on gold colloid/cysteine/nafion-modified platinum disk electrode. *Sensors and Actuators B: Chemical, 115*(1), 109–115.

Liu, Z., Jian, Z., Fang, J., Xu, X., Zhu, X., & Wu, S. (2012). Low-Temperature Reverse Microemulsion Synthesis, Characterization, and Photocatalytic Performance of Nanocrystalline Titanium Dioxide. *International Journal of Photoenergy, 2012.* doi: 10.1155/2012/702503

Lo, Y. L., & Hwang, B. J. (1995). Kinetics of Ethanol Oxidation on Electroless Ni-P/SnO$_2$/Ti Electrodes in KOH Solutions. *Journal of the Electrochemical Society, 142*(2), 445–450. doi: 10.1149/1.2044058

Lou, X., Detrembleur, C., Pagnoulle, C., Jérôme, R., Bocharova, V., Kiriy, A., & Stamm, M. (2004). Surface modification of multiwalled carbon nanotubes by poly (2-vinylpyridine): Dispersion, selective deposition, and decoration of the nanotubes. *Advanced Materials, 16*(23–24), 2123–2127.

Lu, G., Jiang, L., Song, F., & Liu, C. (2005). Determination of Uric Acid and Norepinephrine by Chitosan-Multiwall Carbon Nanotube Modified Electrode. *Electroanalysis, 17*(10), 901–905.

Luo, L., Zhu, L., & Wang, Z. Nonenzymatic amperometric determination of glucose by CuO nanocubes–graphene nanocomposite modified electrode. *Bioelectrochemistry*(0). doi: 10.1016/j.bioelechem.2012.03.006

Luo, M.-l., Tang, W., Zhao, J.-q., & Pu, C.-s. (2006). Hydrophilic modification of poly(ether sulfone) used TiO$_2$ nanoparticles by a sol-gel process. *Journal of Materials Processing Technology, 172*(3), 431–436. doi: 10.1016/j.jmatprotec.2005.11.004

Luo, M. Z., & Baldwin, R. P. (1995). Characterization of carbohydrate oxidation at copper electrodes. *Journal of Electroanalytical Chemistry, 387*(1–2), 87–94. doi: 10.1016/0022-0728(95)03867-g

Luo, X. L., Xu, J. J., Wang, J. L., & Chen, H. Y. (2005). Electrochemically deposited nanocomposite of chitosan and carbon nanotubes for biosensor application. *Chemical Communications*(16), 2169–2171.

Lyons, M. E. G., & Keeley, G. P. (2006). The redox behaviour of randomly dispersed single walled carbon nanotubes both in the absence and in the presence of adsorbed glucose oxidase. *Sensors, 6*(12), 1791–1826.

Macdonald, J. R., & Johnson, W. B. (2005). *Fundamentals of impedance spectroscopy*: Wiley Online Library.

Male, K. B., Hrapovic, S., Liu, Y., Wang, D., & Luong, J. H. T. (2004). Electrochemical detection of carbohydrates using copper nanoparticles and carbon nanotubes. *Analytica Chimica Acta, 516*(1–2), 35–41. doi: 10.1016/j.aca.2004.03.075

Mandizadeh, S., Soofivand, F., Salavati-Niasari, M., & Bagheri, S. (2014). Auto-combustion preparation and characterization of BaFe 12 O 19 nanostructures by using maleic acid as fuel. Journal of Industrial and Engineering Chemistry.

Manjari Lal, V. C., Pushan Ayyub & Amarnath Maitra (1998). Preparation and characterization of ultrafine TiO$_2$ particles in reverse micelles by hydrolysis of titanium di-ethylhexyl sulfosuccinate. *Journal of Materials Research, 13*, 1249–1254 doi: 10.1557/JMR.1998.0178

Mao, S., Li, W., Long, Y., Tu, Y., & Deng, A. (2012). Sensitive electrochemical sensor of tryptophan based on Ag@C core–shell nanocomposite modified glassy carbon electrode. *Analytica Chimica Acta, 738*(0), 35–40. doi: 10.1016/j.aca.2012.06.008

Marcus, R. A. (1964). Chemical and electrochemical electron-transfer theory. *Annual Review of Physical Chemistry, 15*(1), 155–196.

Matsumoto, F., Harada, M., Koura, N., & Uesugi, S. (2003). Electrochemical oxidation of glucose at Hg adatom-modified Au electrode in alkaline aqueous solution. *Electrochemistry Communications, 5*(1), 42–46. doi: 10.1016/s1388-2481(02)00529-5

Mazloum-Ardakani, M., Beitollahi, H., Taleat, Z., Naeimi, H., & Taghavinia, N. (2010). Selective voltammetric determination of d-penicillamine in the presence of tryptophan at a modified carbon paste electrode incorporating TiO_2 nanoparticles and quinizarine. *Journal of Electroanalytical Chemistry, 644*(1), 1–6. doi: 10.1016/j.jelechem.2010.02.034

Meister, A. (1957). Biochemistry of the amino acids. *Biochemistry of the amino acids.*

Menczel, J. D., & Prime, R. B. (2009). *Thermal analysis of polymers*: Wiley Online Library.

Meng, L., Jin, J., Yang, G., Lu, T., Zhang, H., & Cai, C. (2009). Nonenzymatic Electrochemical Detection of Glucose Based on Palladium–Single-Walled Carbon Nanotube Hybrid Nanostructures. *Analytical Chemistry, 81*(17), 7271–7280. doi: 10.1021/ac901005p

Meulenkamp, E. A. (1998). Synthesis and growth of ZnO nanoparticles. *The Journal of Physical Chemistry B, 102*(29), 5566–5572.

Mho, S.-i., & Johnson, D. C. (2001). Electrocatalytic response of carbohydrates at copper-alloy electrodes. *Journal of Electroanalytical Chemistry, 500*(1–2), 524–532. doi: 10.1016/s0022-0728(00)00277-1

Mohanty, P., Kim, B., & J. Park. (2007). Synthesis of single crystalline europium-doped ZnO nanowires. *Materials Science and Engineering: A 138*(3), 224–227. doi: 10.1016/j.mseb.2007.01.007

Moniruzzaman, M., & Winey, K. I. (2006). Polymer nanocomposites containing carbon nanotubes. *Macromolecules, 39*(16), 5194–5205.

Muhd Julkapli, N., Bagheri, S., & Bee Abd Hamid, S. (2014). Recent advances in heterogeneous photocatalytic decolorization of synthetic dyes. The Scientific World Journal, 2014.

Murray, R. W. (1980). Chemically modified electrodes. *Accounts of Chemical Research, 13*(5), 135–141.

Murray, R. W., Goodenough, J., Albery, W., & Murray, R. (1981). Modified Electrodes: Chemically Modified Electrodes for Electrocatalysis [and Discussion]. *Philosophical Transactions of the Royal Society of London. Series A, Mathematical and Physical Sciences, 302*(1468), 253–265.

Musameh, M., Wang, J., Merkoci, A., & Lin, Y. (2002). Low-potential stable NADH detection at carbon-nanotube-modified glassy carbon electrodes. *Electrochemistry Communications, 4*(10), 743–746. doi: 10.1016/s1388-2481(02)00451-4

Nahir, T. M., & Buck, R. P. (1992). Modified Cottrell behavior in thin layers: Applied voltage steps under diffusion control for constant-resistance systems. *Journal of Electroanalytical Chemistry, 341*(1), 1–14.

Narang, J., Chauhan, N., Jain, P., & Pundir, C. S. (2012). Silver nanoparticles/multiwalled carbon nanotube/polyaniline film for amperometric glutathione biosensor. *International Journal of Biological Macromolecules, 50*(3), 672–678. doi: 10.1016/j.ijbiomac.2012.01.023

Nau, V., & Nieman, T. A. (1979). Application of microporous membranes to chemiluminescence analysis. *Analytical Chemistry, 51*(3), 424–428. doi: 10.1021/ac50039a024

Ndamanisha, J. C., & Guo, L. (2009). Nonenzymatic glucose detection at ordered mesoporous carbon modified electrode. *Bioelectrochemistry, 77*(1), 60–63. doi: 10.1016/j.bioelechem.2009.05.003

Newman, J. D., & Turner, A. P. F. (2005). Home blood glucose biosensors: a commercial perspective. *Biosensors and Bioelectronics, 20*(12), 2435–2453. doi: 10.1016/j.bios.2004.11.012

Nie, F.-Q., Xu, Z.-K., Yang, Q., Wu, J., & Wan, L.-S. (2004). Surface modification of poly(acrylonitrile-co-maleic acid) membranes by the immobilization of poly(ethylene glycol). *Journal of Membrane Science, 235*(1–2), 147–155. doi: 10.1016/j.memsci.2004.02.006

Niederberger, M. (2007). Nonaqueous sol-gel routes to metal oxide nanoparticles. *Accounts of Chemical Research, 40*(9), 793–800.

Nieh, T. G., & Wadsworth, J. (1990). Superelastic behaviour of a fine-grained, yttria-stabilized, tetragonal zirconia polycrystal (Y-TZP). *Acta Metallurgica et Materialia, 38*(6), 1121–1133. doi: 10.1016/0956-7151(90)90185-j

Noren, D., & Hoffman, M. (2005). Clarifying the Butler–Volmer equation and related approximations for calculating activation losses in solid oxide fuel cell models. *Journal of Power Sources, 152,* 175–181.

Ohsaka, T., Izumi, F., & Fujiki, Y. (2005). Raman spectrum of anatase, TiO_2. *Journal of Raman Spectroscopy, 7*(6), 321–324.

Ojani, R., Raoof, J.-B., & Norouzi, B. (2009). Electropolymerization of <i> N-methylaniline in the presence of sodium dodecylsulfate and its application for electrocatalytic reduction of nitrite. *Journal of Materials Science, 44*(15), 4095–4103. doi: 10.1007/s10853-009-3591-8

Ojani, R., Raoof, J.-B., & Norouzi, B. (2011). Performance of glucose electrooxidation on Ni–Co composition dispersed on the poly(isonicotinic acid) (SDS) film. *Journal of Solid State Electrochemistry, 15*(6), 1139–1147. doi: 10.1007/s10008-010-1175-9

Özcan, A., & Şahin, Y. (2012). A novel approach for the selective determination of tryptophan in blood serum in the presence of tyrosine based on the electrochemical reduction of oxidation product of tryptophan formed in situ on graphite electrode. *Biosensors and Bioelectronics, 31*(1), 26–31. doi: 10.1016/j.bios.2011.09.048

Park, S., Huh, J. O., Kim, N. G., Kang, S. M., Lee, K.-B., Hong, S. P., ... Choi, I. S. (2008). Photophysical properties of noncovalently functionalized multi-walled carbon nanotubes with poly-para-hydroxystyrene. *Carbon, 46*(4), 714–716. doi:10.1016/j.carbon.2008.01.002

Parpot, P., Nunes, N., & Bettencourt, A. P. (2006). Electrocatalytic oxidation of monosaccharides on gold electrode in alkaline medium: Structure–reactivity relationship. *Journal of Electroanalytical Chemistry, 596*(1), 65–73. doi: 10.1016/j.jelechem.2006.07.006

Patolsky, F., Zayats, M., Katz, E., & Willner, I. (1999). Precipitation of an insoluble product on enzyme monolayer electrodes for biosensor applications: characterization by faradaic impedance spectroscopy, cyclic voltammetry, and microgravimetric quartz crystal microbalance analyses. *Analytical Chemistry, 71*(15), 3171–3180.

Pauliukaite, R., Ghica, M. E., Fatibello-Filho, O., & Brett, C. M. A. (2009). Comparative Study of Different Cross-Linking Agents for the Immobilization of Functionalized Carbon Nanotubes within a Chitosan Film Supported on a Graphite–Epoxy Composite Electrode. *Analytical Chemistry, 81*(13), 5364–5372. doi: 10.1021/ac900464z

Philip, M. (2001). Nanostructured materials. *Reports on Progress in Physics, 64*(3), 297.

Pierre, A. C. (1998). *Introduction to sol-gel processing* (Vol. 1): Springer.

Pournaghi-Azar, M. H., & Ojani, R. (1999). Attempt to incorporate ferrocenecarboxylic acid into polypyrrole during the electropolymerization of pyrrole in chloroform: its application to the electrocatalytic oxidation of ascorbic acid. *Journal of Solid State Electrochemistry, 3*(7), 392–396. doi: 10.1007/s100080050172

Pournaghi-Azar, M. H., & Ojani, R. (2000). Electrochemistry and electrocatalytic activity of polypyrrole/ferrocyanide films on a glassy carbon electrode. *Journal of Solid State Electrochemistry, 4*(2), 75–79. doi: 10.1007/s100080050004

Prakash, T., Navaneethan, M., Archana, J., Ponnusamy, S., Muthamizhchelvan, C., & Hayakawa, Y. (2012). Synthesis of TiO_2 nanoparticles with mesoporous spherical morphology by a wet chemical method. *Materials Letters, 82*(0), 208–210. doi: 10.1016/j.matlet.2012.05.064

Pratsinis, S. E., & Spicer, P. T. (1998). Competition between gas phase and surface oxidation of $TiCl_4$ during synthesis of TiO_2 particles. *Chemical Engineering Science, 53*(10), 1861–1868.

Proença, L., Lopes, M. I. S., Fonseca, I., Kokoh, K. B., Léger, J. M., & Lamy, C. (1997). Electrocatalytic oxidation of d-sorbitol on platinum in acid medium: analysis of the reaction products. *Journal of Electroanalytical Chemistry, 432*(1–2), 237–242. doi: 10.1016/s0022-0728(97)00221-0

Puddu, V., Mokaya, R., & Puma, G. L. (2007). Novel one step hydrothermal synthesis of TiO_2/WO3 nanocomposites with enhanced photocatalytic activity. *Chemical Communications*(45), 4749–4751.

Pumera, M., Merkoçi, A., & Alegret, S. (2007). Carbon nanotube detectors for microchip CE: Comparative study of single-wall and multiwall carbon nanotube, and graphite powder films on glassy carbon, gold, and platinum electrode surfaces. *Electrophoresis, 28*(8), 1274–1280. doi: 10.1002/elps.200600632

Qian, L., & Yang, X. (2006). Composite film of carbon nanotubes and chitosan for preparation of amperometric hydrogen peroxide biosensor. *Talanta, 68*(3), 721–727. doi: 10.1016/j.talanta.2005.05.030

Ramimoghadam, D., Bagheri, S., & Abd Hamid, S. B. (2014a). Biotemplated synthesis of anatase titanium dioxide nanoparticles via lignocellulosic waste material. BioMed research international, 2014.

Ramimoghadam, D., Bagheri, S., & Hamid, S. B. A. (2014b). Progress in electrochemical synthesis of magnetic iron oxide nanoparticles. Journal of Magnetism and Magnetic Materials, 368, 207–229.

Ramimoghadam, D., Bagheri, S., & Hamid, S. B. A. (2015). In-situ precipitation of ultra-stable nano-magnetite slurry. Journal of Magnetism and Magnetic Materials, 379, 74–79.

Rani, S., Suri, P., Shishodia, P., & Mehra, R. (2008). Synthesis of nanocrystalline ZnO powder via sol-gel route for dye-sensitized solar cells. *Solar Energy Materials and Solar Cells, 92*(12), 1639–1645.

Ranjit, K., & Viswanathan, B. (1997). Synthesis, characterization and photocatalytic properties of iron-doped TiO_2 catalysts. *Journal of Photochemistry and Photobiology A: Chemistry, 108*(1), 79–84.

Rao, A. V. P., Robin, A. I., & Komarneni, S. (1996). Bismuth titanate from nanocomposite and sol-gel processes. *Materials Letters, 28*(4–6), 469–473. doi: 10.1016/0167-577x(96)00107-3

Raoof, J.-B., Ojani, R., & Baghayeri, M. (2009). Simultaneous electrochemical determination of glutathione and tryptophan on a nano-TiO_2/ferrocene carboxylic acid modified carbon paste electrode. *Sensors and Actuators B: Chemical, 143*(1), 261–269. doi: 10.1016/j.snb.2009.08.046

Raoof, J.-B., Ojani, R., & Chekin, F. (2007). Electrochemical Analysis of D-Penicillamine Using a Carbon Paste Electrode Modified with Ferrocene Carboxylic Acid. *Electroanalysis, 19*(18), 1883–1889. doi: 10.1002/elan.200703947

Raoof, J. B., Ojani, R., & Rashid-Nadimi, S. (2004). Preparation of polypyrrole/ferrocyanide films modified carbon paste electrode and its application on the electrocatalytic determination of ascorbic acid. *Electrochimica Acta, 49*(2), 271–280.

Reddy, S. M., & Vadgama, P. M. (1997). A study of the permeability properties of surfactant modified poly(vinyl chloride) membranes. *Analytica Chimica Acta, 350*(1–2), 67–76. doi: 10.1016/s0003-2670(97)00314-0

Rotello, V. M. (2004). *Nanoparticles: building blocks for nanotechnology*: Springer.

Rubtsova, M. Y., Kovba, G. V., & Egorov, A. M. (1998). Chemiluminescent biosensors based on porous supports with immobilized peroxidase. *Biosensors and Bioelectronics, 13*(1), 75–85. doi: 10.1016/s0956-5663(97)00072-9

Saetre, R., & Rabenstein, D. L. (1978). Determination of penicillamine in blood and urine by high performance liquid chromatography. *Analytical Chemistry, 50*(2), 276–280. doi: 10.1021/ac50024a027

Safavi, A., Maleki, N., Moradlou, O., & Tajabadi, F. (2006). Simultaneous determination of dopamine, ascorbic acid, and uric acid using carbon ionic liquid electrode. *Analytical Biochemistry, 359*(2), 224–229.

Saidman, S. B., Lobo-Castañón, M. J., Miranda-Ordieres, A. J., & Tuñón-Blanco, P. (2000). Amperometric detection of d-sorbitol with NAD+-d-sorbitol dehydrogenase modified carbon paste electrode. *Analytica Chimica Acta, 424*(1), 45–50. doi: 10.1016/s0003-2670(00)01140-5

Sakaguchi, T., Morioka, Y., Yamasaki, M., Iwanaga, J., Beppu, K., Maeda, H., . . . Tamiya, E. (2007). Rapid and onsite BOD sensing system using luminous bacterial cells-immobilized chip. *Biosensors and Bioelectronics, 22*(7), 1345–1350.

Sanchez, C., Livage, J., Henry, M., & Babonneau, F. (1988). Chemical modification of alkoxide precursors. *Journal of Non-Crystalline Solids, 100*(1–3), 65–76. doi: 10.1016/0022-3093(88)90007-5

Santos, L. M., & Baldwin, R. P. (1987). Liquid chromatography/electrochemical detection of carbohydrates at a cobalt phthalocyanine containing chemically modified electrode. *Analytical Chemistry, 59*(14), 1766–1770. doi: 10.1021/ac00141a006

Sattarahmady, N., Heli, H., & Faramarzi, F. (2010). Nickel oxide nanotubes-carbon microparticles/ Nafion nanocomposite for the electrooxidation and sensitive detection of metformin. *Talanta, 82*(4), 1126–1135. doi: 10.1016/j.talanta.2010.06.022

Schmidt Martin, K. R., von Issendorff Bernd, Haberland Hellmut. (1998). Irregular variations in the melting point of size-selected atomic clusters. *Nature, 393*(6682), 238–240.

Scott, K. (1991). *Electrochemical reaction engineering*: Academic Press.

Segarra Guerrero, R., Sagrado Vives, S., & Martinez Calatayud, J. (1991). Fluorimetric determination of captopril by flow injection analysis. *Microchemical Journal, 43*(3), 176–180. doi: 10.1016/s0026-265x(10)80002-5

Shahrokhian, S., & Bozorgzadeh, S. (2006). Electrochemical oxidation of dopamine in the presence of sulfhydryl compounds: Application to the square-wave voltammetric detection of penicillamine and cysteine. *Electrochimica Acta, 51*(20), 4271–4276. doi: 10.1016/j.electacta.2005.12.006

Shahrokhian, S., & Fotouhi, L. (2007). Carbon paste electrode incorporating multi-walled carbon nanotube/cobalt salophen for sensitive voltammetric determination of tryptophan. *Sensors and Actuators B: Chemical, 123*(2), 942–949. doi: 10.1016/j.snb.2006.10.053

Shameli, K., Ahmad, M. B., Al-Mulla, E. A. J., Shabanzadeh, P., & Bagheri, S. (2015). Antibacterial effect of silver nanoparticles on talc composites. *Research on Chemical Intermediates, 41*(1), 251–263.

Shameli, K., Bin Ahmad, M., Jaffar Al-Mulla, E. A., Ibrahim, N. A., Shabanzadeh, P., Rustaiyan, A., *et al.* (2012). Green biosynthesis of silver nanoparticles using Callicarpa maingayi stem bark extraction. *Molecules, 17*(7), 8506–8517.

Shamsipur, M., Najafi, M., & Hosseini, M.-R. M. (2010). Highly improved electrooxidation of glucose at a nickel(II) oxide/multi-walled carbon nanotube modified glassy carbon electrode. *Bioelectrochemistry, 77*(2), 120–124. doi: 10.1016/j.bioelechem.2009.07.007

Shan, D., He, Y., Wang, S., Xue, H., & Zheng, H. (2006). A porous poly(acrylonitrile-co-acrylic acid) film-based glucose biosensor constructed by electrochemical entrapment. *Analytical Biochemistry, 356*(2), 215–221. doi: 10.1016/j.ab.2006.06.005

Sharma, D., Sharma, S., Kaith, B. S., Rajput, J., & Kaur, M. (2011). Synthesis of ZnO nanoparticles using surfactant free in-air and microwave method. *Applied Surface Science, 257*(22), 9661–9672. doi: 10.1016/j.apsusc.2011.06.094

Sheng, Q., Luo, K., Li, L., & Zheng, J. (2009). Direct electrochemistry of glucose oxidase immobilized on NdPO4 nanoparticles/chitosan composite film on glassy carbon electrodes and its biosensing application. *Bioelectrochemistry, 74*(2), 246–253. doi: 10.1016/j.bioelechem.2008.08.007

Shie, J. W., Yogeswaran, U., & Chen, S. M. (2008). Electroanalytical properties of cytochrome-c by direct electrochemistry on multi-walled carbon nanotubes incorporated with DNA biocomposite film. *Talanta, 74*(5), 1659–1669.

Shoji, E., & Freund, M. S. (2001). Potentiometric Sensors Based on the Inductive Effect on the pKa of Poly(aniline): A Nonenzymatic Glucose Sensor. *Journal of the American Chemical Society, 123*(14), 3383–3384. doi: 10.1021/ja005906j

Siegel, R. W., & Fougere, G. E. (1995). *Grain size dependent mechanical properties in nanophase materials*.

Siesler, H. W., Ozaki, Y., Kawata, S., & Heise, H. M. (2008). *Near-infrared spectroscopy: Principles, instruments, applications*: Wiley-Vch.

Simoyi, M. F., Falkenstein, E., Van Dyke, K., Blemings, K. P., & Klandorf, H. (2003). Allantoin, the oxidation product of uric acid is present in chicken and turkey plasma. *Comparative Biochemistry and Physiology Part B: Biochemistry and Molecular Biology, 135*(2), 325–335.

Singh, R., Verma, R., Sumana, G., Srivastava, A. K., Sood, S., Gupta, R. K., & Malhotra, B. D. (2012). Nanobiocomposite platform based on polyaniline-iron oxide-carbon nanotubes for bacterial detection. *Bioelectrochemistry, 86*(0), 30–37. doi: 10.1016/j.bioelechem.2012.01.005

Smith, B. C. (2009). *Fundamentals of FourierTransform Infrared Spectroscopy*: CRC.

Smyth, M. R., & Vos, J. G. (1992). *Analytical voltammetry* (Vol. 27): Elsevier.

Spinks, G. M., Wallace, G. G., Fifield, L. S., Dalton, L. R., Mazzoldi, A., De Rossi, D., . . . Baughman, R. H. (2002). Pneumatic Carbon Nanotube Actuators. *Advanced Materials, 14*(23), 1728–1732. doi: 10.1002/1521-4095(20021203)14:23<1728::aid-adma1728>3.0.co;2-8

Star, A., Steuerman, D. W., Heath, J. R., & Stoddart, J. F. (2002). Starched carbon nanotubes. *Angewandte Chemie International Edition, 41*(14), 2508–2512.

Stradiotto, N. R., Yamanaka, H., & Zanoni, M. V. B. (2003). Electrochemical sensors: a powerful tool in analytical chemistry. *Journal of the Brazilian Chemical Society, 14*(2), 159–173.

Subrahmanyam, A., Biju, K. P., Rajesh, P., Jagadeesh Kumar, K., & Raveendra Kiran, M. (2012). Surface modification of sol gel TiO_2 surface with sputtered metallic silver for Sun light photocatalytic activity: Initial studies. *Solar Energy Materials and Solar Cells, 101*(0), 241–248. doi: 10.1016/j.solmat.2012.01.023

Suliman, F. E. O., Al-Lawati, H. A. J., Al-Kindy, S. M. Z., Nour, I. E. M., & Salama, S. B. (2003). A sequential injection spectrophotometric method for the determination of penicillamine in pharmaceutical products by complexation with iron(III) in acidic media. *Talanta, 61*(2), 221–231. doi: 10.1016/s0039-9140(03)00250-9

Suryanarayana, C., & Norton, M. G. (1998). *X-ray diffraction: a practical approach*: Springer.

Suzuki, Y., & Yoshikawa, S. (2004). Synthesis and thermal analyses of TiO_2-derived nanotubes prepared by the hydrothermal method. *Journal of Materials Research, 19*(04), 982–985.

Svetlicic, V., Schmidt, A. J., & Miller, L. L. (1998). Conductometric sensors based on the hypersensitive response of plasticized polyaniline films to organic vapors. *Chemistry of Materials, 10*(11), 3305–3307.

Tabata, Y., & Ikada, Y. (1998). Protein release from gelatin matrices. *Advanced Drug Delivery Reviews, 31*(3), 287–301. doi: 10.1016/s0169-409x(97)00125-7

Tang, J., Redl, F., Zhu, Y., Siegrist, T., Brus, L. E., & Steigerwald, M. L. (2005). An organometallic synthesis of TiO_2 nanoparticles. *Nano Letters, 5*(3), 543–548.

Tang, X., Liu, Y., Hou, H., & You, T. (2010). Electrochemical determination of L-Tryptophan, L-Tyrosine and L-Cysteine using electrospun carbon nanofibers modified electrode. *Talanta, 80*(5), 2182–2186. doi: 10.1016/j.talanta.2009.11.027

Taniguchi, N. (1974). *On the basic concept of nanotechnology*. Paper presented at the Proc. Intl. Conf. Prod. Eng. Tokyo, Part II, Japan Society of Precision Engineering.

Tasviri, M., Rafiee-Pour, H. A., Ghourchian, H., & Gholami, M. R. (2011). Amine functionalized TiO_2 coated on carbon nanotube as a nanomaterial for direct electrochemistry of glucose oxidase and glucose biosensing. *Journal of Molecular Catalysis B-Enzymatic, 68*(2), 206–210. doi: 10.1016/j.molcatb.2010.11.005

Termeh Yousefi, A., Bagheri, S., Shinji, K., Rusop Mahmood, M., & Ikeda, S. (2014). Highly oriented vertically aligned carbon nanotubes via chemical vapour deposition for key potential application in CNT ropes. Materials Research Innovations.

TermehYousefi, A., Bagheri, S., & Adib, N. (2015a). Integration of biosensors based on microfluidic: a review. Sensor Review, 35(2), 190–199.

Termehyousefi, A., Bagheri, S., Kadri, N., Elfghi, F. M., Rusop, M., & Ikeda, S. (2015b). Synthesis of Well-Crystalline Lattice Carbon Nanotubes via Neutralized Cooling Method. Materials and Manufacturing Processes, 30(1), 59–62.

TermehYousefi, A., Bagheri, S., Kadri, N. A., Mahmood, M. R., & Ikeda, S. (2015c). Constant Glucose Biosensor Based on Vertically Aligned Carbon Nanotube Composites. Int. J. Electrochem. Sci, 10, 4183–4192.

TermehYousefi, A., Bagheri, S., Shinji, K., Rouhi, J., Rusop Mahmood, M., & Ikeda, S. (2014). Fast synthesis of multilayer carbon nanotubes from camphor oil as an energy storage material. BioMed research international, 2014.

TermehYousefi, A., Tanaka, H., Bagheri, S., Elfghi, F., Rusop, M., & Ikeda, S. (2015d). Vectorial Crystal Growth of Oriented Vertically Aligned Carbon Nanotubes using statistical analysis. Crystal Growth & Design.

Tkac, J., Whittaker, J. W., & Ruzgas, T. (2007). The use of single walled carbon nanotubes dispersed in a chitosan matrix for preparation of a galactose biosensor. Biosensors and Bioelectronics, 22(8), 1820–1824. doi: 10.1016/j.bios.2006.08.014

Tlili, A., Abdelghani, A., Ameur, S., & Jaffrezic-Renault, N. (2006). Impedance spectroscopy and affinity measurement of specific antibody–antigen interaction. Materials Science and Engineering: C, 26(2), 546–550.

Toito Suarez, W., Marcolino Jr, L. H., & Fatibello-Filho, O. (2006). Voltammetric determination of N-acetylcysteine using a carbon paste electrode modified with copper(II) hexacyanoferrate(III). Microchemical Journal, 82(2), 163–167. doi: 10.1016/j.microc.2006.01.007

Tokumoto, M. S., Pulcinelli, S. H., Santilli, C. V., & Briois, V. (2003). Catalysis and temperature dependence on the formation of ZnO nanoparticles and of zinc acetate derivatives prepared by the sol-gel route. The Journal of Physical Chemistry B, 107(2), 568–574.

Tominaga, M., Shimazoe, T., Nagashima, M., & Taniguchi, I. (2005). Electrocatalytic oxidation of glucose at gold nanoparticle-modified carbon electrodes in alkaline and neutral solutions. Electrochemistry Communications, 7(2), 189–193. doi: 10.1016/j.elecom.2004.12.006

Torriero, A. A. J., Piola, H. D., Martínez, N. A., Panini, N. V., Raba, J., & Silber, J. J. (2007). Enzymatic oxidation of tert-butylcatechol in the presence of sulfhydryl compounds: Application to the amperometric detection of penicillamine. Talanta, 71(3), 1198–1204. doi: 10.1016/j.talanta.2006.06.027

Torriero, A. A. J., Salinas, E., Marchevsky, E. J., Raba, J., & Silber, J. J. (2006). Penicillamine determination using a tyrosinase micro-rotating biosensor. Analytica Chimica Acta, 580(2), 136–142. doi: 10.1016/j.aca.2006.07.067

Torto, N., Ruzgas, T., & Gorton, L. (1999). Electrochemical oxidation of mono- and disaccharides at fresh as well as oxidized copper electrodes in alkaline media. Journal of Electroanalytical Chemistry, 464(2), 252–258. doi: 10.1016/s0022-0728(99)00041-8

Trentler, T. J., Denler, T. E., Bertone, J. F., Agrawal, A., & Colvin, V. L. (1999). Synthesis of TiO2 nanocrystals by nonhydrolytic solution-based reactions. Journal of the American Chemical Society, 121(7), 1613–1614.

Uguina, M. A., Ovejero, G., Van Grieken, R., Serrano, D. P., & Camacho, M. (1994). Synthesis of Titanium Silicalite-1 from an SiO2-TiO2 Cogel Using a Wetness Impregnation Method. ChemInform, 25(18), no-no.

Upadhyay, S., Bagheri, S., & Hamid, S. B. A. (2014). Enhanced photoelectrochemical response of reduced-graphene oxide/Zn 1- x Ag x O nanocomposite in visible-light region. international journal of hydrogen energy, 39(21), 11027–11034.

Vafaee, M., & Ghamsari, M. S. (2007). Preparation and characterization of ZnO nanoparticles by a novel sol-gel route. Materials Letters, 61(14), 3265–3268.

Vasantha, V., & Chen, S. M. (2006). Electrocatalysis and simultaneous detection of dopamine and ascorbic acid using poly (3, 4-ethylenedioxy) thiophene film modified electrodes. Journal of Electroanalytical Chemistry, 592(1), 77–87.

Vaseashta, A., & Dimova-Malinovska, D. (2005). Nanostructured and nanoscale devices, sensors and detectors. *Science and Technology of Advanced Materials, 6*(3), 312–318.

Vidotti, M., Cerri, C. D., Carvalhal, R. F., Dias, J. C., Mendes, R. K., Córdoba de Torresi, S. I., & Kubota, L. T. (2009). Nickel hydroxide electrodes as amperometric detectors for carbohydrates in flow injection analysis and liquid chromatography. *Journal of Electroanalytical Chemistry, 636*(1–2), 18–23. doi: 10.1016/j.jelechem.2009.09.006

Vieira, D. F., & Pawlicka, A. (2010). Optimization of performances of gelatin/LiBF4-based polymer electrolytes by plasticizing effects. *Electrochimica Acta, 55*(4), 1489–1494. doi: 10.1016/j.electacta.2009.04.039

Vilgis, T. A., Heinrich, G., & Klüppel, M. (2009). *Reinforcement of Polymer Nano-composites: Theory, Experiments and Applications*: Cambridge University Press Cambridge, UK.

Viñas, P., Garcia, I. L., & Gil, J. A. M. (1993). Determination of thiol-containing drugs by chemiluminescence—flow injection analysis. *Journal of Pharmaceutical and Biomedical Analysis, 11*(1), 15–20. doi: 10.1016/0731-7085(93)80144-p

Walter Kochen (Editor), H. S. E. (1994). *L-Tryptophan: Current Prospects in Medicine and Drug Safety*.

Walter Kochen, H. S. (1994). *l-Tryptophan—Current Prospects in Medicine and Drug Safety;*. Berlin: Walter de Gruyter.

Wang, C. C., & Ying, J. Y. (1999). Sol-gel synthesis and hydrothermal processing of anatase and rutile titania nanocrystals. *Chemistry of Materials, 11*(11), 3113–3120.

Wang, J. (1988). *Electroanalytical techniques in clinical chemistry and laboratory medicine*: Vch New York.

Wang, J. (2006). *Analytical electrochemistry*: Wiley-VCH.

Wang, J., Li, M., Shi, Z., Li, N., & Gu, Z. (2002). Direct Electrochemistry of Cytochrome c at a Glassy Carbon Electrode Modified with Single-Wall Carbon Nanotubes. *Analytical Chemistry, 74*(9), 1993–1997. doi: 10.1021/ac010978u

Wang, J., Musameh, M., & Lin, Y. (2003). Solubilization of carbon nanotubes by Nafion toward the preparation of amperometric biosensors. *Journal of the American Chemical Society, 125*(9), 2408–2409.

Wang, J., Shi, N., Qi, Y., & Liu, M. (2010). Reverse micelles template assisted fabrication of ZnO hollow nanospheres and hexagonal microtubes by a novel fast microemulsion-based hydrothermal method. *Journal of Sol-Gel Science and Technology, 53*(1), 101–106. doi: 10.1007/s10971-009-2063-6

Wang, J., & Taha, Z. (1990). Catalytic oxidation and flow detection of carbohydrates at ruthenium dioxide modified electrodes. *Analytical Chemistry, 62*(14), 1413–1416. doi: 10.1021/ac00213a013

Wang, M., Pramoda, K. P., & Goh, S. H. (2006). Enhancement of interfacial adhesion and dynamic mechanical properties of poly(methyl methacrylate)/multiwalled carbon nanotube composites with amine-terminated poly(ethylene oxide). *Carbon, 44*(4), 613–617. doi: 10.1016/j.carbon.2005.10.001

Wang, Q., Dong, D., & Li, N. (2001). Electrochemical response of dopamine at a penicillamine self-assembled gold electrode. *Bioelectrochemistry, 54*(2), 169–175. doi: http://dx.doi.org/10.1016/S1567-5394(01)00125-6

Wang, S. G., Zhang, Q., Wang, R., & Yoon, S. F. (2003). A novel multi-walled carbon nanotube-based biosensor for glucose detection. *Biochemical and Biophysical Research Communications, 311*(3), 572–576. doi: 10.1016/j.bbrc.2003.10.031

Wang, X., Zhuang, J., Peng, Q., & Li, Y. (2005). A general strategy for nanocrystal synthesis. *Nature, 437*(7055), 121–124.

Wang, Z., Liu, J., Liang, Q., Wang, Y., & Luo, G. (2002). Carbon nanotube-modified electrodes for the simultaneous determination of dopamine and ascorbic acid. *Analyst, 127*(5), 653–658.

Wang, Z. L. (2004). Zinc oxide nanostructures: growth, properties and applications. *Journal of Physics: Condensed Matter 16*(25), R829.

Warren, B. E. (1990). *X-ray Diffraction*: Dover publications.

Weetall, H. H. (1974). Immobilized Enzymes: Analytical Applications. *Analytical Chemistry, 46*(7), 602A–615A. doi: 10.1021/ac60343a721

West, R. H., Celnik, M. S., Inderwildi, O. R., Kraft, M., Beran, G. J. O., & Green, W. H. (2007). Toward a comprehensive model of the synthesis of TiO_2 particles from TiCl4. *Industrial & Engineering Chemistry Research, 46*(19), 6147–6156.

Widrig, C. A., Porter, M. D., Ryan, M. D., Strein, T. G., & Ewing, A. G. (1990). Dynamic electrochemistry: methodology and application. *Analytical Chemistry, 62*(12), 1R–20R. doi: 10.1021/ac00211a009

Wightman, R., Wipf, D., & Bard, A. (1989). Electroanalytical Chemistry. *Electroanalytical Chemistry, 15*.

Williams, D. B., & Carter, C. B. (2009). *Transmission electron microscopy: a textbook for materials science*: Springer.

Willner, I., & Katz, E. (2006). *Bioelectronics*: Wiley-VCH.

Wilson, G. S., & Gifford, R. (2005). Biosensors for real-time in vivo measurements. *Biosensors and Bioelectronics, 20*(12), 2388–2403.

Wolfe, R. L., Balasubramanian, R., Tracy, J. B., & Murray, R. W. (2007). Fully ferrocenated hexanethiolate monolayer-protected gold clusters. *Langmuir, 23*(4), 2247–2254.

Wooten, M., & Gorski, W. (2010). Facilitation of NADH Electro-oxidation at Treated Carbon Nanotubes. *Analytical Chemistry, 82*(4), 1299–1304. doi: 10.1021/ac902301b

Wroński, M. (1996). Separation of urinary thiols as tributyltinmercaptides and determination using capillary isotachophoresis. *Journal of Chromatography B: Biomedical Sciences and Applications, 676*(1), 29–34. doi: 10.1016/0378-4347(95)00404-1

Wu, B., Zhang, G., Shuang, S., & Choi, M. M. F. (2004). Biosensors for determination of glucose with glucose oxidase immobilized on an eggshell membrane. *Talanta, 64*(2), 546–553. doi: 10.1016/j.talanta.2004.03.050

Wu, F. H., Zhao, G. C., Wei, X. W., & Yang, Z. S. (2004). Electrocatalysis of tryptophan at multi-walled carbon nanotube modified electrode. *Microchimica Acta, 144*(4), 243–247.

Wu, L., Zhang, X., & Ju, H. (2007). Amperometric glucose sensor based on catalytic reduction of dissolved oxygen at soluble carbon nanofiber. *Biosensors and Bioelectronics, 23*(4), 479–484. doi: 10.1016/j.bios.2007.06.009

Wu, X., Zhao, F., Varcoe, J. R., Thumser, A. E., Avignone-Rossa, C., & Slade, R. C. T. (2009). Direct electron transfer of glucose oxidase immobilized in an ionic liquid reconstituted cellulose–carbon nanotube matrix. *Bioelectrochemistry, 77*(1), 64–68. doi: 10.1016/j.bioelechem.2009.05.008

Xie, Y., Qian, H., Zhong, Y., Guo, H., & Hu, Y. (2012). Facile Low-Temperature Synthesis of Carbon Nanotube/TiO_2 Nanohybrids with Enhanced Visible-Light-Driven Photocatalytic Activity. *International Journal of Photoenergy, 2012*. doi: 10.1155/2012/682138

Xu, J., Wang, Y., Xian, Y., Jin, L., & Tanaka, K. (2003). Preparation of multiwall carbon nanotubes film modified electrode and its application to simultaneous determination of oxidizable amino acids in ion chromatography. *Talanta, 60*(6), 1123–1130.

Yang, L., Xiong, H., Zhang, X., Wang, S., & Zhang, X. (2011). Direct electrochemistry of glucose oxidase and biosensing for glucose based on boron-doped carbon-coated nickel modified electrode. *Biosensors and Bioelectronics, 26*(9), 3801–3805. doi: 10.1016/j.bios.2011.02.037

Ye, J., & Baldwin, R. P. (1994). Determination of carbohydrates, sugar acids and alditols by capillary electrophoresis and electrochemical detection at a copper electrode. *Journal of Chromatography A, 687*(1), 141–148.

Ye, J. S., Wen, Y., De Zhang, W., Ming Gan, L., Xu, G. Q., & Sheu, F. S. (2004). Nonenzymatic glucose detection using multi-walled carbon nanotube electrodes. *Electrochemistry Communications, 6*(1), 66–70.

Yeo, I.-H., & Johnson, D. C. (2001). Electrochemical response of small organic molecules at nickel–copper alloy electrodes. *Journal of Electroanalytical Chemistry, 495*(2), 110–119. doi: 10.1016/s0022-0728(00)00401-0

Yildiz, H. B., Kiralp, S., Toppare, L., & Yağci, Y. (2005). Immobilization of glucose oxidase in conducting graft copolymers and determination of glucose amount in orange juices with enzyme electrodes. *International Journal of Biological Macromolecules, 37*(4), 174–178. doi: 10.1016/j.ijbiomac.2005.10.004

Yin, H., Wada, Y., Kitamura, T., Kambe, S., Murasawa, S., Mori, H., … Yanagida, S. (2001). Hydrothermal synthesis of nanosized anatase and rutile TiO_2 using amorphous phase TiO_2. *Journal of Materials Chemistry, 11*(6), 1694–1703.

Ying, I.., Kang, E. T., & Neoh, K. G. (2002). Covalent immobilization of glucose oxidase on microporous membranes prepared from poly(vinylidene fluoride) with grafted poly(acrylic acid) side chains. *Journal of Membrane Science, 208*(1–2), 361–374. doi: 10.1016/s0376-7388(02)00325-3

Yogeswaran, U., & Chen, S. M. (2007a). Electrocatalytic properties of electrodes which are functionalized with composite films of f-MWCNTs incorporated with poly (neutral red). *Journal of the Electrochemical Society, 154*(11), E178–E186.

Yogeswaran, U., & Chen, S. M. (2007b). Separation and concentration effect of f-MWCNTs on electrocatalytic responses of ascorbic acid, dopamine and uric acid at f-MWCNTs incorporated with poly (neutral red) composite films. *Electrochimica Acta, 52*(19), 5985–5996.

Yogeswaran, U., & Chen, S. M. (2008). Multi-walled carbon nanotubes with poly (methylene blue) composite film for the enhancement and separation of electroanalytical responses of catecholamine and ascorbic acid. *Sensors and Actuators B: Chemical, 130*(2), 739–749.

Yogeswaran, U., Thiagarajan, S., & Chen, S. M. (2007a). Nanocomposite of functionalized multiwall carbon nanotubes with nafion, nano platinum, and nano gold biosensing film for simultaneous determination of ascorbic acid, epinephrine, and uric acid. *Analytical Biochemistry, 365*(1), 122–131.

Yogeswaran, U., Thiagarajan, S., & Chen, S. M. (2007b). Pinecone shape hydroxypropyl-β-cyclodextrin on a film of multi-walled carbon nanotubes coated with gold particles for the simultaneous determination of tyrosine, guanine, adenine and thymine. *Carbon, 45*(14), 2783–2796.

Yousefi, A. T., Ikeda, S., Mahmood, M. R., Rouhi, J., & Yousefi, H. T. (2012). Encapsulation Technology Based on Coacervation Method to Control Release of Manufactured Nano Particles. World Applied Sciences Journal, 17(4), 524–531.

Yousefi, A. T., Ikeda, S., Rusop Mahmood, M., & Yousefi, H. T. (2014). Simulation of Nano Sensor Based on Carbon Nanostructures in Order to Form Multifunctional Delivery Platforms. Advanced Materials Research, 832, 778–782.

You, T., Niwa, O., Chen, Z., Hayashi, K., Tomita, M., & Hirono, S. (2003). An Amperometric Detector Formed of Highly Dispersed Ni Nanoparticles Embedded in a Graphite-like Carbon Film Electrode for Sugar Determination. *Analytical Chemistry, 75*(19), 5191–5196. doi: 10.1021/ac034204k

Yu, H., Yu, J., Cheng, B., & Zhou, M. (2006). Effects of hydrothermal post-treatment on microstructures and morphology of titanate nanoribbons. *Journal of Solid State Chemistry, 179*(2), 349–354.

Yu, J. C., Zhang, L., Zheng, Z., & Zhao, J. (2003). Synthesis and Characterization of Phosphated Mesoporous Titanium Dioxide with High Photocatalytic Activity. *Chemistry of Materials, 15*(11), 2280–2286. doi: 10.1021/cm0340781

Zadeii, J. M., Marioli, J., & Kuwana, T. (1991). Electrochemical detector for liquid chromatographic determination of carbohydrates. *Analytical Chemistry, 63*(6), 649–653. doi: 10.1021/ac00006a019

Zak, A. K., Majid, W. H. A., Darroudi, M., & Yousefi, R. (2011). Synthesis and characterization of ZnO nanoparticles prepared in gelatin media. *Materials Letters, 65*(1), 70–73. doi: 10.1016/j.matlet.2010.09.029

Zaleska, A. (2008). Doped-TiO$_2$: A Review. *Recent Patents on Engineering, 2*(3), 157–164. doi: 10.2174/187221208786306289

Zamiri, G., Bagheri, S., Shahnazar, S., & Hamid, S. (2015). Progress on synthesis, functionalisation and applications of graphene nanoplatelets. Materials Research Innovations, 1433075X1433015Y. 0000000044.

Zangmeister, R. A., Park, J. J., Rubloff, G. W., & Tarlov, M. J. (2006). Electrochemical study of chitosan films deposited from solution at reducing potentials. *Electrochimica Acta, 51*(25), 5324–5333. doi: 10.1016/j.electacta.2006.02.003

Zhai, H.-J., Wu, W.-H., Lu, F., Wang, H.-S., & Wang, C. (2008). Effects of ammonia and cetyltrimethylammonium bromide (CTAB) on morphologies of ZnO nano- and micromaterials under solvothermal process. *Materials Chemistry and Physics, 112*(3), 1024–1028. doi: 10.1016/j.matchemphys.2008.07.020

Zhan, G. D., Kuntz, J. D., Wan, J., & Mukherjee, A. K. (2003). Single-wall carbon nanotubes as attractive toughening agents in alumina-based nanocomposites. *Nature Materials, 2*(1), 38–42.

Zhang, H., Chen, B., Banfield, J. F., & Waychunas, G. A. (2008). Atomic structure of nanometer-sized amorphous TiO$_2$. *Physical Review B, 78*(21), 214106.

Zhang, H., Meng, Z., Wang, Q., & Zheng, J. (2011). A novel glucose biosensor based on direct electrochemistry of glucose oxidase incorporated in biomediated gold nanoparticles–carbon nanotubes composite film. *Sensors and Actuators B: Chemical, 158*(1), 23–27. doi: 10.1016/j.snb.2011.04.057

Zhang, H., Yang, D., Ji, Y., Ma, X., Xu, J., & Que, D. (2004). Low temperature synthesis of flowerlike ZnO nanostructures by cetyltrimethylammonium bromide-assisted hydrothermal process. *The Journal of Physical Chemistry B, 108*(13), 3955–3958.

Zhang, M., Bando, Y., & Wada, K. (2001). Sol-gel template preparation of TiO$_2$ nanotubes and nanorods. *Journal of materials science letters, 20*(2), 167–170.

Zhang, M., Mullens, C., & Gorski, W. (2006). Amperometric glutamate biosensor based on chitosan enzyme film. *Electrochimica Acta, 51*(21), 4528–4532. doi: 10.1016/j.electacta.2006.01.010

Zhang, M., Smith, A., & Gorski, W. (2004). Carbon nanotube-chitosan system for electrochemical sensing based on dehydrogenase enzymes. *Analytical Chemistry, 76*(17), 5045–5050.

Zhang, S., Wang, N., Niu, Y., & Sun, C. (2005). Immobilization of glucose oxidase on gold nanoparticles modified Au electrode for the construction of biosensor. *Sensors and Actuators B: Chemical, 109*(2), 367–374.

Zhang, S., Wang, N., Yu, H., Niu, Y., & Sun, C. (2005). Covalent attachment of glucose oxidase to an Au electrode modified with gold nanoparticles for use as glucose biosensor. *Bioelectrochemistry, 67*(1), 15–22.

Zhang, X., Wang, J., Wang, Z., & Wang, S. (2005). Improvement of amperometric sensor used for determination of nitrate with polypyrrole nanowires modified electrode. *Sensors, 5*(12), 580–593.

Zhang, X. J., Wang, G. F., Huang, Y., Yu, L. T., & Fang, B. (2011). Copper(II) doped nanoporous TiO$_2$ composite based glucose biosensor. *Analytical Methods, 3*(11), 2611–2615. doi: 10.1039/c1ay05368j

Zhao, C., Shao, C., Li, M., & Jiao, K. (2007). Flow-injection analysis of glucose without enzyme based on electrocatalytic oxidation of glucose at a nickel electrode. *Talanta, 71*(4), 1769–1773. doi: 10.1016/j.talanta.2006.08.013

Zhao, G. C., Yin, Z. Z., Zhang, L., & Wei, X. W. (2005). Direct electrochemistry of cytochrome *c* on a multi-walled carbon nanotubes modified electrode and its electrocatalytic activity for the reduction of H_2O_2. *Electrochemistry Communications, 7*(3), 256–260.

Zhao, Q., Guan, L., Gu, Z., & Zhuang, Q. (2004). Determination of phenolic compounds based on the tyrosinase-single walled carbon nanotubes sensor. *Electroanalysis, 17*(1), 85–88.

Zhao, W., Xu, J. J., Shi, C. G., & Chen, H. Y. (2005). Multilayer membranes via layer-by-layer deposition of organic polymer protected Prussian blue nanoparticles and glucose oxidase for glucose biosensing. *Langmuir, 21*(21), 9630–9634.

Zhou, B., Hermans, S., & Somorjai, G. A. (2003). *Nanotechnology in Catalysis Volumes 1 and 2* (Vol. 1): Springer.

Zhou, H., Fan, T., & Zhang, D. (2007). Hydrothermal synthesis of ZnO hollow spheres using spherobacterium as biotemplates. *Microporous and mesoporous materials, 100*(1), 322–327.

Zhou, X., Nie, H., Yao, Z., Dong, Y., Yang, Z., & Huang, S. (2012). Facile synthesis of nanospindle-like Cu2O/straight multi-walled carbon nanotube hybrid nanostructures and their application in enzyme-free glucose sensing. *Sensors and Actuators B: Chemical, 168*(0), 1–7. doi: 10.1016/j.snb.2011.12.012

Zhuo, Y., Yuan, R., Chai, Y., Zhang, Y., Li, X., Wang, N., & Zhu, Q. (2006). Amperometric enzyme immunosensors based on layer-by-layer assembly of gold nanoparticles and thionine on Nafion modified electrode surface for α-1-fetoprotein determinations. *Sensors and Actuators B: Chemical, 114*(2), 631–639.